Which Babies
Shall Live?

Which Babies Shall Live?, edited by **Thomas H. Murray** and **Arthur L. Caplan**, 1985

Feeling Good and Doing Better, edited by **Thomas H. Murray, Willard Gaylin,** and **Ruth Macklin**, 1984

Ethics and Animals, edited by **Harlan B. Miller** and **William H. Williams**, 1983

Profits and Professions, edited by **Wade L. Robison, Michael S. Pritchard,** and **Joseph Ellin**, 1983

Visions of Women, edited by **Linda A. Bell**, 1983

Medical Genetics Casebook, by **Colleen Clements**, 1982

Who Decides?, edited by **Nora K. Bell**, 1982

The Custom-Made Child?, edited by **Helen B. Holmes, Betty B. Hoskins,** and **Michael Gross**, 1981

Birth Control and Controlling Birth, edited by **Helen B. Holmes, Betty B. Hoskins,** and **Michael Gross**, 1980

Medical Responsibility, edited by **Wade L. Robison** and **Michael S. Pritchard**, 1979

Contemporary Issues in Biomedical Ethics, edited by **John W. Davis, Barry Hoffmaster,** and **Sarah Shorten**, 1979

Which Babies Shall Live?

Humanistic Dimensions
of the Care of
Imperiled Newborns

Edited by

*Thomas H. Murray
and Arthur L. Caplan*

Humana Press • Clifton, New Jersey

Library of Congress Cataloging in Publication Data
Main entry under title:

Which babies shall live?

   (Contemporary issues in biomedicine, ethics,
and society)
   "Product of the Hastings Center's Research
Group on Ethics and the Care of Newborns
project"—Acknowledgments.
   Includes bibliographies and index.
   1. Infants (Newborn)—Diseases—Treatment—
Moral and ethical aspects.   2. Euthanasia—
Moral and ethical aspects.   3. Infanticide—
Moral and ethical aspects.   I. Murray, Thomas H.
II. Caplan, Arthur L.   III. Hastings Center.
Research Group on Ethics and the Care of New—
borns.   IV. Series.
RJ255.W49   1985   362.1'989201   85–18058
ISBN 0–89603–086–5

©1985 The Humana Press Inc.
Crescent Manor
PO Box 2148
Clifton, NJ 07015

Printed in the United States of America

Preface

The fate of seriously ill newborns has captured the attention of the public, of national and state legislators, and of powerful interest groups. For the most part, the debate has been cast in the narrowest possible terms: "discrimination against the handicapped"; "physician authority"; "family autonomy." We believe that something much more profound is happening: the debate over the care of sick and dying babies appears to be both a manifestation of great changes in our feelings about infants, children, and families, and a reflection of deep and abiding attitudes toward the newborn, the handicapped, and perhaps other humans who are "less than" normal, rational adults.

How could we cast some light on those feelings and attitudes that seemed to determine silently the course of the public debate? We chose to enlist the humanities—the displayers and critics of our cultural forms. Rather than closing down the public discussion, we wanted to open it up, to illuminate it with the light of history, religion, philosophy, literature, jurisprudence, and humanistically oriented sociology. This book is a first effort to place the hotly contested Baby Doe debate into a broader cultural context.

We expect this book to appeal to all those interested in the debate over our treatment of seriously ill newborns. Nurses, physicians, and social service professionals, along with those interested in the public policy debate—lawyers, people in government, and concerned laypersons—join philosophers and other humanists as the primary audience for *Which Babies Shall Live?*. We do not expect that this book will "solve" the problem of ill newborns. We do believe, however, that the perspectives it brings will deepen the reader's understanding

v

of the issues, and generate a healthy respect for their complexity and their cultural roots.

Following an Introduction that summarizes the key public cases in which a decision was made not to treat a newborn, the book is organized into four sections. The first section, "The Child, Medicine, and Science," sets the stage with a neonatologist's description of the history and current state of neonatology. Next an historian of youth and the family examines the long-standing tension between scientific judgments of what is "best" for children, and the resistance to those judgments. An historian and a lawyer close the section with their critical commentaries.

The next section, "Religion, Suffering, and Morality," begins with a theologian's discussion of the ways our religious traditions have regarded infants. Then comes a chapter that reflects on the prevalence of suffering in neonatal intensive care, including the suffering of parents and caretakers, as well as the infants, and considers the moral significance of suffering in that context. A philosopher in the next chapter subjects the widely favored "best interest of the infant" principle to careful scrutiny, and shows how our belief that we know what the principle means or how it can be applied may be ill-founded. Finally, a lawyer who directs an "ethics consult team" in a hospital discusses privacy as a consideration in the refusal of care for newborns.

A provocative literary critic and author opens the third section, "Images of the Abandoned," with a plea to respect the diversity of human forms, especially "abnormality" in its many guises. Through his study of "freaks," the author concludes that our rejection of humans who look or act differently from ourselves may rob us of the richness of variations that are a source of human strength. This section ends with a rejoinder by a philosopher.

The fourth section, "Caretakers and Attitudes," reports the results of a survey taken at the symposium for which the other papers were originally composed. The survey posed many of the central questions about what principles should govern decisions about ill newborns, and who should make those decisions. For the first time, detailed information is available about the beliefs and attitudes of those involved in caring for these babies, including nurses and physicians. The volume ends with a chapter that sums up the main themes of the book.

Which Babies Shall Live? is unique in its approach. It is thoroughly interdisciplinary, it concentrates on the contributions of the Humanities, and it refrains from offering "solutions" where none are available. It is also written to be understandable by intelligent people who are not specialists in any of the disciplines on display.

We fervently hope that readers of this book will come away with an enriched understanding of the cultural context in which the Babies Doe have been born, and some, at least, have died. We further hope that readers will see that the humanities are indispensable to a full understanding of problems like the care of imperiled infants, and that the perspectives offered by the humanities will help us to appreciate the richness and depth of other problems as well.

Thomas H. Murray
Arthur L. Caplan

Acknowledgments

This volume is the first book-length product of The Hastings Center's Research Group on Ethics and the Care of Newborns project, with the support of the March of Dimes, Upjohn Company, Squibb, and the Pew Memorial Trust. The Research Group's work continues, and more publications are expected, but we owe special thanks to the individuals and institutions who aided us in our early attempts to identify and clarify ethical issues in the care of imperiled newborns.

The Montefiore Hospital and Medical Center and our colleagues there, John Arras, Nancy Dubler, and Alan Fleischman, deserve a special thank you: the individuals, for their advice and unstinting support; the Medical Center for its co-sponsorship of the conference on which this book is based. Also, we would like to thank the administration at the Benjamin F. Cardozo School of Law, which graciously permitted us the use of not one, but two auditoriums for the day of the program, and especially Alan Weisbard, lawyer, friend, critic, and advocate for the conference.

Someone had to pay for all this, and in this instance the credit (or rather the debits) went to the New York Council for the Humanities. We thank our Program Officer, Edward Bristow, and the Director, Jay Kaplan, for their encouragement.

Lastly, we must thank our coworkers at The Hastings Center, especially Mary Gualandi, Eva Mannheimer, and Eric Feldman, without whose efforts we would have ended up having our speakers meet at a corner booth at the Empire Diner with three people who wandered in off the street.

CONTENTS

Religion, Suffering, and Morality

Images of the Abandoned

Caretakers: Images and Attitudes

Contributors

JOHN D. ARRAS • *Department of Epidemiology and Social Medicine, Albert Einstein College of Medicine, Montefiore Hospital and Medical Center, Bronx, NY*

ARTHUR L. CAPLAN • *The Hastings Center, Hastings-on-Hudson, NY*

NANCY N. DUBLER • *Department of Social Medicine, Montefiore Hospital and Medical Center, Bronx, NY*

LESLIE A. FIEDLER • *Department of English, SUNY-Buffalo, Buffalo, NY*

ALAN R. FLEISCHMAN • *Montefiore Hospital and Medical Center, Bronx, NY*

JOSEPH F. KETT • *Department of History, University of Virginia, Charlottesville, VA*

BETTY WOLDER LEVIN • *Division of Sociomedical Sciences, Columbia School of Public Health, New York, NY*

RUTH MACKLIN • *Department of Community Health, Albert Einstein College of Medicine, Bronx, NY*

THOMAS H. MURRAY • *Institute for the Medical Humanities, The University of Texas Medical Branch, Galveston, TX*

DAVID H. SMITH • *Department of Religious Studies, Indiana University, Bloomington, IN*

MARGARET O'BRIEN STEINFELS • *National Pastoral Life Center, New York, NY*

ALAN J. WEISBARD • *Cardozo School of Law, Yeshiva University, New York, NY*

Beyond Babies Doe

An Introduction

Thomas H. Murray and Arthur L. Caplan

We are trapped in a difficult scene of our own construction. Public debate over the care of seriously ill newborns has cast the issue in a stark light, a light that obliterates the persistent complexities and special circumstances that everywhere tend to render simplistic solutions unworkable and, ultimately, painful to someone—parents, doctors, nurses, or the babies themselves.

In the midst of an intricate struggle among the federal government, physicians, hospitals, and parents, there is little room for uncertainty, ambivalence, and honest confusion. Victory belongs to those who are full of conviction and quick spoken, even though their solutions may be ill-conceived and inappropriate.

Especially at a time like this, we can benefit by taking a vacation, however brief, from pretending to know the answers, and reflect for a while about those deeper currents that prompt us to think and feel and react in the ways we do about the plight of those babies whose futures rest in our hands. If we turn away, even for a moment, from statutes and regulations—which are only the superficial manifestations of our ethical convictions, tempered by practicality—we can look to the sources of our moral and legal life. The deepest insights into the underground river of our ethos seem to be those provided by the humanities, the disciplines that best reflect and amplify the often inchoate concerns and percep-

1

tions exhibited in our literature, religion, philosophy, and culture.

We see our ethical problems not as we might view some abstract set of statistics, but instead as concrete, embodied images. When we talk about caring for imperiled newborns, we tend not to think about the most common sort of case—the very premature infant—but about a baby with Down's syndrome who also has a life-threatening problem, or one who has spina bifida, hydrocephalus and microcephaly. Little wonder then that the law chose to think of nontreatment decisions of such infants as "discrimination against the handicapped." The plights of these infants do not raise any more complex or difficult moral issues than those of very premature babies, and as difficult cases, they are still less common. Is it reasonable to imagine that laws and regulations drawn up with one kind of case in mind may be inappropriate when applied to very different cases? Once we have identified the specific concerns and images that moved us to action, we can begin to inquire into their source and power, and the ways they may aid us, or mislead us.

The articles in this book were called into being for a conference held at the Cardozo Law School in New York City that was sponsored by The Hastings Center, The Montefiore Hospital and Medical Center, and the New York Council for the Humanities. We believed that the humanities had something to offer to the people who occupy the trenches in the battle against infant death and disability—that is, to nurses and physicians—as well as others concerned with moral issues in newborn care. We called on scholars in literature, religion, ethics, history, and jurisprudence, asking them to speak on matters pertinent to our feelings, beliefs, and actions towards infants and children. To our amazement, the expected audience of about a hundred swelled to almost three hundred. The speakers that day struck a uniform note of excellence, prompting us to preserve a record of what was said in this volume.

Before taking up the questions raised by each of the authors, it might be useful to review how the debate over the care of newborns in general, and nontreatment in particular, evolved toward its current state. The best way to do that is to look at three prominent cases of proposed nontreatment, and at the reaction of the federal government and others toward the decisions that were taken in each case. The three cases are the Siamese twins in Danville, Illinois, Baby Doe in Bloomington, Indiana, and Infant Jane Doe, in Stony Brook, New York.

The Danville, Illinois Twins

Male twins were born on May 5, 1981 in Danville, Illinois. The parents and their doctors were expecting twins, but they did not anticipate that the boys' bodies would be joined in the abdominal region as Siamese twins. The twins share a lower stomach and bowel, and have three legs. According to one account, which cannot be confirmed, the obstetrician overseeing the birth ordered that the babies not be resuscitated. Nonetheless the children lived. By May 13, the Illinois Department of Children and Family Services had taken temporary custody of the twins, alleging that the parents and physicians had ordered that the children be given only palliative medical care, but not food or water. The babies were moved to a major medical center nearby, Children's Memorial Hospital in Chicago.

The case took an unprecedented turn when the Vermilion County district attorney filed criminal charges against the parents and physicians involved in the case. The charges included attempted homicide and conspiracy to commit homicide. This appears to be the first time criminal charges were made in a case of nontreatment for a newborn.

Throughout the incident, no one claimed that the parents were callous, or were doing what they believed was wrong for their children. A nurse at the hospital in which the children were born described in court the mother's reaction: "She [the mother] said after seeing the X-rays, seeing that the twins shared a common pelvis, bowel and stomach, she was sure they were doing the right thing."[1] On the contrary, the parents appeared to be very compassionate. Nurses testified that the mother visited the boys often, cuddled them, and showed deep emotional attachment to them.

The criminal charges got nowhere. At a preliminary hearing on July 17 held to see if there was probable cause that those charged had committed a crime, no witness would testify that the parents or physicians had ordered that the infants not be fed. The charges were dismissed for lack of evidence.

On September 16, 1981, four months after the babies' birth, they were returned to the custody of their parents.

We can ask what might move parents or physicians to decide in favor of an early death for babies like these. We cannot know what motives were present in the Danville case, and should not judge the participants. But we may learn something by looking at the possible reasons that could have been offered.

One unpleasant and unlikely scenario could have been embarrassment and even horror on the part of the parents, who saw in the babies' speedy death the end of a terrible episode. In this version, the children's welfare is not even considered; only the effects on the parents are acknowledged.

Much more likely is a version in which the parents and physicians made their decision on what they believed were the babies' best interests. It would have been a plausible response, even if wrong. Perhaps those present felt that the babies, cyanotic at birth and so physically deformed, would die soon in any case. In that instance, no good would be done them by merely prolonging their inevitable death. We know now that the infants survived; thus, if that had been the reason for not treating or feeding them, it would have been factually incorrect and, had it not been overruled, a tragic and irreversible mistake.

Or, rather than thinking that the children were already dying, perhaps the physicians and parents feared that, though the babies might live, their lives would be filled with suffering—physical, psychological, or both. As Thomas Murray's essay reveals, the concern for suffering has importance for both parents and health care providers, although their understanding of this concept does not always overlap.

The physical appearance of the children may have played a crucial role in influencing the decisions of the parents. Their appearance might have suggested that they would need constant medical attention to survive. Or, most likely, the adults present at the birth simply reacted with an instinctive horror to the physical malformations, and judged that others would do likewise. From there, it is a short step to inferring that a life in which others would treat one as an object of revulsion—or, worse, amusement—would not be worth living. Thus, it might have been a desire to spare the children that sort of life-long emotional trauma that prompted the choice to permit their death.

Advocates for the disabled tell us it is difficult enough to lead a satisfying life with a physical disability without having people judge that such a life is worse than death. And, as Leslie Fiedler points out in his chapter in this book, the record does not bear out the presumption that physically anomalous people lead miserable lives. On the contrary, people with disabilities can and do lead lives of intermingled joy and sorrow, just as those who are "normal" do. Our projections of misery may have more to do with our own fears than with the reality of life with a malformed or malfunctioning body.

There are two other lessons lurking in the Danville Siamese twin case. First, as the details of the case were gradually revealed, people seemed to place great importance on the question of whether their brains were normal or damaged. When it was learned that the brain of one, and possibly both children, was probably healthy, this information seemed to solidify support in favor of saving the children. This brought home the distinction we intuitively make between physical and mental abnormalities, a distinction given fatal force in the case which follows, Baby Doe of Bloomington.

The other lesson is about our profound uncertainty over what is to be done in such cases. As Joseph Kett's paper reveals, such ambivalence has long been present in Americans' reactions to their children. Public policy with respect to the medical treatment of children has been as much a function of economics and professional skill as it has of moral or religious outlook. Moreover, differences of opinion between health providers, legal authorities and parents have been and continue to be common in this area.

Some nurses felt strongly enough about those children's right to be fed and to live that they apparently violated the physician's orders, fed them, and reported the situation to the state. As Betty Levin's paper shows, such feelings are quite common among nurses asked to carry out a physician's order not to feed or provide a child with fluids. Yet, no one felt strongly enough that the parents and physicians had done something culpably wrong and deserved to be punished that they were willing to testify against them in court. This compromise—let the babies live but punish no one for the decision to let them die—may be Solomonic in its wisdom, especially in the face of great social and moral confusion over the moral status of handicaps, nontreatment decisions, and the authority of parents over their children. Before condemning anyone, we should remember the nature and extent of current disagreements and confusion.

Baby Doe, Bloomington, Indiana

What the Danville twins did to raise ethical questions about physical abnormalities, Baby Doe of Bloomington did for mental ones. Baby Doe was born at Bloomington Hospital on April 9, 1982. The infant experienced some distress during delivery, and was limp and cyanotic at birth, although he was more responsive a few minutes later. He showed obvious evidence of Down's syndrome, a probable tracheoesophageal fistula (a hole connecting the trachea and the esophagus), and a possible heart anomaly.[2] On the advice of the obstetrician who had delivered the infant, the parents chose not to allow an operation to repair the tracheoesophageal fistula. The child was sedated and given pain medication, and died six days later.

Legal efforts to override the parents' and doctor's decision in Indiana courts failed; the child died before the case could be brought to a federal court. But the case of Baby Doe transformed the way decisions on whether to treat seriously ill newborns are handled in the United States. The US Department of Health and Human Services rapidly set up a toll-free "hotline" on which instances of nontreatment could be reported. The Department claimed that Section 504 of the Rehabilitation Act of 1973 gave it authority to investigate cases in which seriously ill infants were denied treatment and, if it was determined that discrimination against the handicapped was involved, to cut off federal funds to that institution. DHHS issued a new regulation to this effect, which was then quickly thrown out by Judge Gerhardt Gesell, principally on the grounds that DHHS had not followed the proper procedures in issuing the rule. In particular, there had been no provision for public comment on its content. DHHS then issued a revised rule, with a comment period. On January 12, 1984, a final rule was issued, reasserting the Department's right to investigate these cases, and to have immediate access to hospital records.[3] The final rule does encourage hospitals to establish "Infant Care Review Committees." These committees would review suggested nontreatment decisions for seriously ill newborns, but they have only as much authority as DHHS chooses to give them. The Department promises to try to contact the appropriate committee when it contemplates launching an investigation, but it retains full authority to overrule or sidestep the committees.

The case of Baby Doe is doubly significant. For one thing, it poses with crystalline clarity the question of whether mental retardation ought to be considered in making a treatment decision. Surgery to correct a tracheoesophageal fistula is complex, but is done successfully in about 90% of attempts. No one seriously disputes the conclusion that, if Baby Doe had not had the extra chromosome that made it a Down's syndrome rather than a mentally normal child, surgery certainly would have been tried. It was the retardation caused by Down's syndrome that made the difference, a fact identified and condemned in numerous editorials, including those appearing in *The New York Times* and the *Washington Post*.[4]

For another, Baby Doe became a dominant image in the public debate over nontreatment for seriously ill newborns. Since April 1982, to discuss nontreatment evokes images of starving to death a retarded infant whose life could probably have been saved. Paradoxically, the prevailing sentiment in the bioethics community seems to be that Baby Doe should have been treated, and that Down's syndrome is no justification to withhold lifesaving treatment. The federal rule uses the term "handicapped," which does apply to Baby Doe, as if it covered all the kinds of cases in which nontreatment might be considered. There are so many more problematic cases, in which genuine moral dilemmas do exist; yet, when we discuss the possibility of justifiable nontreatment, the specter of Baby Doe looms unavoidably, as an examplar of wrongful nontreatment. And if all cases were like that of Baby Doe, finding the morally defensible answer would not be so difficult. But, sadly, not all cases are like Baby Doe, as Infant Jane Doe demonstrated.

Infant Jane Doe

As clearly as the case of Baby Doe allows us to make a moral judgment that he should have been treated, the case of Infant Jane Doe seems mired in factual and moral confusion. Infant Jane Doe was born on October 11, 1983, in Port Jefferson, New York. Early reports were vague about her condition, although all seemed to agree that she had at least three problems: myelomeningocele, a condition in which the spinal column fails to cover the spinal cord, which lies exposed on the back with its surrounding membranes; hydrocephalus, where the fluid normally produced in the brain cannot drain off, with consequent buildup of pressure on brain tissue; and microcephaly, an abnormally small head. An opinion filed by the US Court of Appeals for the Second Circuit on February 23, 1984, lists other birth defects including a "weak face"—the infant cannot close her eyes or suck properly; a malformed brainstem; spasticity of the upper extremities; and a thumb entirely within her fist.[5]

She was quickly transferred to University Hospital at the State University of New York at Stony Brook, where her parents and physician decided not to perform surgery to close the lesion on her back. This type of surgery has two main purposes: first, it minimizes further damage to the spinal cord, and second, it greatly decreases the chance that the child will contract meningitis—an infection of the cerebrospinal fluid with a potentially devastating impact on the brain.[6] The parents argued that Infant Jane Doe was already so damaged that surgery would merely prolong a pain-filled, burdensome life, and that it would be more compassionate to allow the child to die sooner rather than later.

On October 16, 1983, a Vermont attorney, A. Lawrence Washburn, Jr., petitioned the New York State Supreme Court (despite the name "Supreme," the State's lowest court) to appoint a guardian *ad litem* for the baby, and to order that the surgery be performed. Judge Melvyn Tannenbaum conducted a hearing on October 19 and 20, at the end of which he appointed William E. Weber, a local attorney, as guardian *ad litem* and authorized him to consent to the surgery, on the child's behalf.

The next day, October 21, the Appellate Division of the New York State Supreme Court (an intermediate level) reversed Judge Tannenbaum's ruling, on the grounds that "[t]hese concerned and loving parents have made an informed, intelligent, and reasonable determination based upon and supported by responsible medical authority."[7]

The case reached the State's highest court, the New York Court of Appeals, almost immediately. On October 28, the Court of Appeals ruled in favor of the hospital and parents, and against Weber. In harsh language, the court singled out attorney Washburn for criticism, and said generally, "[i]t would serve no useful purpose at this stage to recite the unusual, and sometimes offensive, activities and proceedings of those who have sought at various stages, in the interests of Baby Jane Doe, to displace parental responsibility for and management of her medical care."[8]

The court also expressed concern that permitting intervention of a third party with no relationship to the child or the child's care, and no authority from the State "would catapult him into the very heart of a family circle, there to challenge the most private and most precious responsibility vested in the parents for the care and nurture of their children . . ."[9]

On another front, the United States Department of Health and Human Services, which had mysteriously received Infant Jane Doe's medical records for the first eight days of her life, on October 22, 1983 made the first of numerous attempts to obtain the child's medical records subsequent to the first eight days. University Hospital refused, and DHHS filed a suit in the US District Court for the Eastern District of New York, alleging that the hospital was violating section 504 of the Rehabilitation Act of 1973, the crux of the "Baby Doe" Rule. Judge Leonard D. Wexler heard the case. During oral arguments on November 17, the federal lawyers conceded that no evidence of discrimination could be found in the records of the child's first eight days. Judge Wexler dismissed the suit on several grounds, and added "the decision of the parents to refuse consent to the surgical procedures was a reasonable one based on due consideration of the medical options available and on a genuine concern for the best interests of the child."[10]

This standard—the best interests of the child—has emerged as the one most commentators, including David H. Smith in his discussion of religious perspectives on the treatment of infants, and the editors of this volume, support as the proper one to use. For a number of reasons it seems far more practical to attempt to determine what would be in a particular child's best interest than it would be to attempt to ask someone to try and make a judgment from the point of view of the child as to what it would or would not want with respect to treatment. Substituted judgment, a doctrine that has been invoked in many controversial cases regarding treatment for older patients, is very difficult to apply to those who have never expressed their views on any subject. However, there are numerous problems associated with a "best interests" standard as well. Just how ambiguous a "best interests" standard can be in certain circumstances is tellingly illustrated by the article by John Arras in this volume.

The federal government did not give up, but instead appealed the case to the US Court of Appeals for the Second Circuit. On February 23, 1984, that court handed down its verdict, very general in scope. The court relied heavily on a review of the history of the legislation on which the government asserted its authority to intervene in Infant Doe cases. They claim their review "has shown that congress never contemplated that section 504 of the Rehabilitation Act would apply to treatment decisions involving defective newborn infants when the statute was enacted in 1973, when it was amended in 1974, or at any subsequent time. Further, neither the articulated purposes of the statute, which concern access and admission to federally-funded programs and activities for otherwise qualified handicapped individuals, nor any fair and reasonable projections of those purposes . . . nor the applicable case law interpreting the statute, support the far-reaching position advanced by the government in this case."[11] In other words, the US Court of Appeals has called into question the rationale used by the federal government for intervening in these cases at all. They take one step further and urge that Congress, not the Executive Branch, make the decision about how to balance the various interests at stake in cases like Infant Jane Doe.

In the midst of the legal wrangling the government again urged that hospitals consider the desirability of formulating committees to review treatment decisions regarding newborns. How the government came to mention review committees at all is an interesting story in itself.

The idea of involving an "ethics" committee when nontreatment was being considered for newborns had been around for a while, but it was taken seriously when the American Academy of Pediatrics (AAP) proposed "Infant Bioethical Review Committees" in its official response to the proposed federal rule.[12] The lack of enthusiasm in DHHS's reception of the idea is indicated by the negligible authority the final rule gives the committees. Even more interesting is what happened to ethics when the AAP proposal was translated into the federal rule: the name changed from "Infant *Bioethical* Review Committee" to "Infant *Care* Review Committee," "care" having a stronger medical than ethical connotation, and the expert on ethical issues was dropped from the list of suggested members. The Department apparently felt uncomfortable with acknowledging the ethical dimensions of these decisions.

This collection has the modest goal of attempting to shed some light on the images that control our perceptions of and responses to dilemmas in newborn care. Until we sort them out much better than we have done so far, we will be battling with the shadows of our own minds and hearts. We owe it to the parents, nurses, and doctors, and, most of all, to the

babies, to confront our own demons so that we may deal
wisely and compassionately with the very real tragedies they
confront every day.

Notes and References

[1]Joyce Wadler, "Deformed Siamese Twins: Should They Live, Who Should
    Decide?" *Washington Post*, June 5, 1981, p. A5.
[2]John E. Pless, "The Story of Baby Doe," *New England Journal of Medicine*,
    Sept. 5, 1983, 309, No. 11, p. 664.
[3]Department of Health and Human Services, 45 CFR Part 84,
    Nondiscrimination on the Basis of Handicap: Procedures and Guide-
    lines Relating to Health Care for Handicapped Infants," *Federal Regis-
    ter*, Jan. 12, 1984, **49,** No. 8, pp. 1622–1654.
[4]*NY Times*, " 'Private' Death," April 26, 1982; *Washington Post*, "The
    Bloomington Baby," April 20, 1982.
[5]US Court of Appeals for the Second Circuit No. 679, August Term, 1983,
    Decided February 23, 1984; Docket No. 83-6343. *US of America v.
    University Hospital, State of New York at Stony Brook.*
[6]Anthony Gallo, "Spina Bifida: The State of the Art of Medical Manage-
    ment," *Hastings Center Report*, 14, No. 1, Feb. 1984, pp. 10–13.
[7]*William E. Weber v. Stony Brook Hospital et al.*, Appellate Division, Second
    Judicial Department, State of New York.
[8]Decision of NY. Court of Appeals in case of *Weber v. Stony Brook Hospital, et
    al.*, October 28, 1983. The court based its ruling on the grounds that
    there are legally mandated procedures for reviewing cases like Infant
    Jane Doe through the Family Court Act, and attorneys Weber and
    Washburn attempted to circumvent those procedures, p. 11.
[9]Ibid.
[10]Opinion of Judge Leonard D. Wexler, quoted in United States Court of
    Appeals for the Second Circuit, No. 679, August Term, 1983. Docket
    No. 83–6343, *United States of America v. University Hospital, State
    University of New York and Parent of Baby Jane Doe*, pp. 1938–1939.
[11]US Court of Appeals.
[12]American Academy of Pediatrics, "Comments of the AAP on Proposed
    Rule Regarding Nondiscrimination on the Basis of Handicap Relat-
    ing to Health Care for Handicapped Newborn Infants," American
    Academy of Pediatrics, July 5, 1983.

Section I

The Child, Medicine, and Science

Caring for Babies in Danger

The Evolution and Current State of Neonatology

Alan R. Fleischman

Introduction

The sickest patients in our medical centers are frequently also the smallest. In no area of medicine are the ethical dilemmas imposed by an exploding technology more poignant than in Perinatology. Physicians, nurses, midwives, and parents responsible for the care of the pregnant woman and the newborn have always realized that there were ethical dilemmas inherent in the very nature of their work. In the last 20 years, with the increase in interest and resources, the care for the fetus and newborn has developed into a new and exciting field.

Perinatology is the care of the pregnant woman and her fetus. Neonatology begins with concern for the fetus and continues into the care of the newborn infant. It is a natural offspring of its parent field, Pediatrics, but depends on input from its colleagues in Obstetrics. The greatest risks for future morbidity and mortality of any time during life occur within the first 24 hours after birth. The neonatologist has accepted the responsibility for guiding the infant through this period of transition to extrauterine independent life. The development of that new human being into a functioning member of society is our ultimate concern. The neonatologist deals with a unique patient who may as yet be unborn, is always unknowing about the outside world, unable to partake in decision-making, and potentially unable even to care for itself. The neonatologist becomes the advocate and consultant as well as the caregiver for the patient.

History of Neonatology

The field of neonatology has a short history, beginning around the turn of this century when it was first appreciated that temperature regulation played a significant role in the survival of small infants. Infants less than four pounds survived at a rate four times greater if their rectal temperature could be maintained in the normal range of 35 to 37° centigrade.[1] This led to the design of the incubator, the first significant contribution to the field of newborn medicine. At the turn of the century there were no premature units in hospitals, and exhibits were held at fairs and amusement parks at which people paid money to view the premature babies in their little incubators.

In 1922, in Chicago, the first hospital-based premature unit was opened.[2] Concern quickly focused on methods of feeding infants who were too small, sick, or weak to be maintained with breast feeding. Since breast feeding was the sole source of food for newborns prior to this century, artificial formulas had to be developed when breast milk could not be utilized. Throughout the 1930s, 40s, and 50s extensive investigations on artificial formulas were conducted and many such formulas were created, some of which still exist today. Investigative methods were devised to determine requirements for the care of newborns in terms of their nutrition, their fluids, and special problems in their medications. Subsequent to World War II, when resources could be directed toward health care delivery and preventive health, neonatology flourished and care for the sick newborn became commonplace. In the 1960s, with the advent of intensive care for adults and a concern for heart attack and trauma victims, some of these techniques began to be utilized in the care of newborns.

The 1960s and 70s saw the logarithmic growth of neonatal intensive care, bringing the new technology of respirators, careful monitoring, and aggressive intervention to the sick newborn.

Current Neonatology

Approximately 250,000 of the 3,500,000 babies born in the United States each year will require neonatal intensive care. This care ranges from a few hours of observation to many weeks and months of intensive care, depending upon the level of illness in the patient. In an affluent middle-class community hospital approximately three to four percent of babies will require some special care or attention. In an inner city, urban, lower socioeconomic setting, as many as 14 to 15% of babies will require special care because of many maternal and environmental influences. There are approximately 600 hospitals providing 7500 neonatal intensive care beds in the United States, and about 1200 pediatricians who are subspecialty trained specifically in neonatology. Although pediatricians in general know how to care for newborns and infants, the care for these very sick patients is reserved to the subspecialists in the neonatal centers.

The cost has been estimated at $8000 to $100,000 per baby for a total of anywhere from 1.5 to 3 billion dollars per year.[3] Many feel that the cost of this care is excessive; this of course depends upon one's perspective. The expenditure might be compared to the cost for one MX missile or, alternatively, compared to the cost of an extensive program for the prevention of premature delivery in a population of urban poor.

There has been a great deal written about the expectations of the general public for survival of newborns in our neonatal intensive care units. Virtually weekly, if not daily, newspaper articles, radio, television, and other media, highlight the "miracles" of neonatology, giving the public an expectation that any pregnancy cared for with prenatal care can and will result in a newborn who is normal and without problem. In fact, over the last 50 years there has been an increasing survival rate for smaller and sicker newborns, and new thresholds of viability continue to be broken on a yearly basis. It is no longer acceptable, however, to define the success of neonatology only in terms of mortality and survival. The more important question is the quality of that survival. There are data from our center and others supporting the view that the quality of survival is increasing as the overall survival rates are increasing. However, there continues to be a substantial number of newborns who are surviving with handicaps, some of them mild and some of them dramatically severe.

Types of Neonatal Intensive Care Patients

The patients being cared for in our neonatal intensive care units can be categorized into four groups: first are the prematures, those babies born many weeks before they are due and with weights that are far below that of a full-term baby; second are the babies with congenital anomalies, defects, and malformations that can be minor in physical importance or major, and other defects that can be minor in cosmetic importance or major; the third group are the babies with genetic disorders, that is, disorders in chromosomal complement or makeup. The Down's Syndrome baby falls into this group, but so do babies with far more severe mental retardation, congenital anomalies, and other defects that are genetically determined. These babies include newborns with trisomies 13 and 18 and other rare genetic defects; the fourth group of newborns being cared for in our neonatal units are the asphyxiated babies. These are full-term or premature infants who have had some problem that has caused a lack of oxygen flow to their brains that results in serious brain malfunction or brain damage, as well as the potential for damage to other vital organs such as the heart, liver, and kidneys. These asphyxiated newborns are frequently resuscitated, placed on ventilators, and aggressively treated, the extent of brain damage becoming obvious only after some period of time in the nursery.

Types of Ethical Dilemmas in Neonatal Care

The information that we have presented is merely an introduction to the myriad of ethical dilemmas that we face in the nursery. These ethical issues can be categorized into two groups: the substantive issues and the procedural ones. The substantive ones are those in which we ask the question "What is right?" The procedural issues are those in which we ask "Who should decide?"

At the Montefiore Medical Center and the North Central Bronx Hospital of the Albert Einstein College of Medicine for the last seven years we have been engaged in exploring these two relatively difficult, but extremely important, areas of ethics with a program we call Neonatal Ethics Rounds.[4] Trained philosophers join the neonatal faculty, fellows, nursing staff, social workers, residents, and other professionals involved in the nursery setting, to discuss a current case, twice a month. These rounds have not been used for decision-making, but rather for education and discussion in which all can explore the bases for making critical decisions within the neonatal unit.

Substantive Issues

From these rounds and from the care of sick newborns and their families, we have learned that the substantive issues include the following questions: Are there newborns who should not be resuscitated? Does initial resuscitation mandate continuation of care? Can essential care ever be withheld? Is there a quality of life below which life is not worth living? Is there a difference between ordinary and extraordinary care? Are economic or psychosocial issues relevant in decision-making? Is there a moral difference between passive and active euthanasia? Are medicolegal concerns relevant to ethical decision-making?

Procedural Issues

The procedural issue of "who should decide" has proven to be inseparable from the substantive issues. Depending upon the answers to the substantive questions raised, it becomes clear in nearly every case who should make the decision for the newborn. Most people inherently believe it should be the parents. They are said to have an "identity of interests" with the newborn and to be closer than anyone else to being an advocate for this sick patient.[5] Furthermore, they will have the responsibility for the care of the newborn after hospitalization and for the rest of its dependent life. However, it should be noted that parents are not unbiased observers. They are often in a state of deep emotional turmoil, concerned about other siblings, fiscal realities, and the stresses of the moment. Furthermore, parents may have certain religious or moral beliefs that may lead to choices not in the best interests of the newborn. For all of these reasons, parents' absolute authority to make decisions relevant to their newborn must be questioned.

Well then, should physicians be allowed to make these decisions? Are they unbiased advocates of the newborn? Certainly, physicians bring to medicine their own biases and feelings and are far from dispassionate observers of their patients' illness. Some have suggested that hospital committees consisting of lay people, social workers, and others be instituted to make some of these decisions. Is this seemingly unbiased advocacy group close enough to the situation to understand the sensitive issues involved? And, finally, many have suggested that only the courts or the government can make these decisions as unbiased representatives of society.

Our Best Answers to Date

We believe that specific answers to the substantive and procedural issues can be found by careful analysis. Answers to the substantive questions can be derived from data presently available concerning prognosis for mortality and morbidity and from a consensus of responsible individuals. These answers can stand as guidelines in individual cases.

We believe that the following are tentative answers to some of the questions: There are newborns who should not be resuscitated. The appropriate limits need to be defined. Initial resuscitation does not mandate continuation of care when new information concerning prognosis and the extent of damage becomes available. As McCormick has suggested, there is a quality of life below which human interaction or relationship is not possible and that quality of life is not worth living.[6] There is sometimes a point in neonatal care where further treatment becomes only a prolongation of pain, suffering, and dying rather than the prolongation of life. At this point, further treatment is inhumane.

All of these substantive answers must be viewed in the light of medical certainty and clear prognostic indicators. When there is uncertainty or a lack of medical facts, the only alternative is aggressive treatment of the neonate.

We further believe that procedures can be evolved, similar to those recommended by the President's Commission for the Study of Ethical Problems, that intimately involve parents and physicians, as well as others within the hospital setting, that can protect against idiosyncratic use of substantive guidelines, and that can most appropriately advocate for the newborn in an environment close enough to the situation to be sensitive, and objective enough to be fair.[7]

We have attempted to present a short review of the field of neonatology and to raise the questions that are troubling those of us who care for critically ill and extremely small babies. It is obvious to us that a compassionate health care delivery system and a supportive society will be necessary in order to resolve the dilemmas associated with providing optimal and appropriate care and caring for our neonates and their families.

References

[1]Dunham, E. C. Evolution of premature infant care. Ann. Pediatr. Fenniae 3:170, 1957.

[2]Hess, J. H. Premature and congenitally diseased infants. Phila, Lea & Febiger, 1922.

[3]Office of Technology Assessment: Cost effectiveness analysis of medical technology—#10, Neonatal Intensive Care, Congress of the United States, Wash, D.C., 1981.

[4]Fleischman, A. Teaching medical ethics in a pediatric training program. Pediatric Annals 10:411, 1981.

[5]Duff, R. and Campbell, A. Moral and ethical dilemmas in the special care nursery. NEJM 289:890, 1973.

[6]McCormick, R. To save or let die—the dilemma of modern medicine. JAMA 229:172, 1974.

[7]Fleischman, A. and Murray, T. Ethics committees for Infants Doe?—An alternative that deserves a chance. Hastings Center Report 13:5, 1983.

Science and Controversy in the History of Infancy in America

Joseph F. Kett

Introduction

Historical studies of childhood tend to be filled with generalizations about momentous changes, yet historical reality often resists generalizations. One problem is that what the intellectuals of a particular period say about children is not necessarily a good description of what their contemporaries do to children. Consider the romantic movement of the late 18th and early 19th centuries as exemplified by William Blake's *Songs of Innocence* (1789) and William Wordsworth's *Ode: Intimations of Immortality from Recollections of Early Childhood* (1807). To these poets, adulthood was a deprivation of the emotional and spiritual response to nature found in children. For romantics, the child was a symbol of creativity, a living witness against social convention. In America we had little Eva and Huck Finn. Yet the good child often became the dead child before the end of the story; think of little Eva and of Dickens' Little Nell. Further, the same century that produced fictional children like these used eight-year-old girls to drag coal through wet tunnels ten hours a day. As John Sommerville notes, "children became an obsessive theme in Victorian culture at the same time that they were being exploited as never before." As symbols of purity in an increasingly ugly world, children were too good to live and hence were quietly put to death by Victorian novelists to punish the cruel adults who had exploited them.[1]

23

Along the same line, consider for a moment the 18th century. The two most prominent English philanthropies of the 18th century were for the benefit of children: the charity school movement sought to educate children of the poor; the London Foundling Hospital, established in 1741 in imitation of one started a century earlier in Paris, provided care for abandoned infants. Surely these were signs of progress. But were they really? The Foundling Hospital graced a city in which, between 1730 and 1750, 75% of those born died by the age of five. Some contemporaries thought that the Foundling Hospital only made things worse. One wrote:

The Hospital Foundling came out of the Brains
To encourage the progress of vulgar amours,
The breeding of rogues and the increasing of Whores,
While children of honest and good husbands and wives,
Stand exposed to oppression and want all their lives.[2]

Contemporaries said much the same about charity schools. They merely made the poor uppity and possibly rebellious. John Trenchard, an English radical Whig from whom American revolutionaries of the 1770s drew enlightened ideas about resistance to oppression, asked "what benefit can accrue to the Publick by taking the Dregs of the People out of the Kennels, and throwing their Betters into them."[3]

Regardless of an age's professions, in sum, the experiences of its children were often harsh. Today we are sensitive to child abuse at least to the extent that we believe that there is such a thing as abuse, and that when it occurs, abusive parents are usually responsible. In the past children did not need to wait on parents for abuse for they were often abused by the daily routine of life. And the abuse was most intense if the child happened to be an infant of a year or two. We are confronted by parents who left their infants alone for long periods, who seem to have been indifferent to their welfare, who could not even remember their names, who refused to attend the funerals of children under five, who routinely farmed infants out for wet nursing, and who argued in divorce proceedings not over which parent could have the infant but over which could unload the infant.[4]

Some of these practices reflected outright malevolence on the part of parents, but more often they resulted from indifference. Indifference reflected, in turn, a certain fatalism about children in general and infants in particular, that is, a belief that there was actually very little that parents could do that would make a difference. This fatalism extended not only to the treatment of infancy but also to childbirth, which most thought was beyond human control. Impregnation, conception, gestation, and parturition were known to be natural processes, but no more under the control of man or woman than were earthquakes, also natural processes. Just as children were gifts from God, monstrosities and infant death were judgments from God. When the 17th-century New England antinomian Anne Hutchinson gave birth to a grapelike cluster of tissues, Puritan authorities in Massachusetts, who had sentenced her to expulsion from the Bay Colony for her heretical views of the workings of divine grace, wanted to know the exact number of lumps in the tissues, in order to ascertain whether they corresponded to the number of her heresies and thus could be construed as a judgment from God.[5]

In 17th century America there was little sense that vital experiences like birth were within human control. The same was true of infancy; God would deal with infants according to His plan. Paradoxically, these same Puritans had an intense interest in education, indeed a keener interest than could be found among their contemporaries almost anywhere in the world.[6] But education could begin only when the child was impressionable, and infants did not appear to be impressionable. From Freud to Watson, the 20th century has emphasized the importance of the first few years of life; in contrast, the 17th century saw little significance in the events of such an early period.

I have emphasized the fatalism that suffused 17th-century attitudes toward infancy in order to call your attention to what strikes me as the major change in the history of infancy, the emergence of a belief that there was a rational, scientific mode of childbirth and infant care. Long before one can point to any real improvement in infant mortality, it is possible to identify the germs of the idea that some modes of treatment were more rationally defensible than others. During the 18th century, both in England and America, a minority of physicians began to argue for a science of infancy. Just as

Newton had demonstrated the laws that governed the workings of the heavens, these men began to suspect that scientific
laws governed parturition. Birth came to be thought of as a
machine-like process in which the womb and birth canal acted
like a mechanical pump. Originating in France in the 17th
century, this view came to affect British medicine by the
mid-18th century. From Britain it crossed the minds of young
American doctors who studied medicine in Edinburgh and
returned to organize the first American medical schools between 1765 and 1800. This movement influenced the rise of
interest in medical care for infants that was evidenced in the
establishment of the Foundling Hospital in London and by
the growing number of treatises on the care of infancy. William Cadogan in England (who launched the Foundling Hospital) and Benjamin Rush in America argued the then
astounding proposition that infant mortality was not inevitable.[7]

No revolution in practice flowed naturally from the work
of these men. Indeed their ideas seem crude to us. An American treatise, published in 1811, urged that mothers put infants before the fire and vigorously rub their limbs, in order
"to give spring to the blood."[8] In fact, in some respects the
new scientific wisdom of the 18th century actually set back the
cause of improved care for infants. The new scientific wisdom
of the 18th century thought little of female midwives. Cadogan, for example, believed that nothing would ever improve
as long as most midwives were women, for women relied on
custom rather than on science. In America, enlightened ideas
about childbirth went hand in hand with an attack on female
midwives and the gradual absorption of midwifery by male
physicians. The prominent role of the latter in childbirth became one of the distinguishing features of 19th-century
medical practice in America. The trouble was that most male
physicians lacked clinical training in dealing with the birth of
children. Their knowledge of childbirth, in other words, came
after, not before, they began practice.[9]

None of this, however, changes the fact that during the
19th century the medical profession insisted that it alone
possessed the science of childbirth. Toward the end of the
century, further, a number of factors led the profession into a

popular health crusade to establish in the public's mind its own preeminent position in matters relating to parturition and infant care. For one thing, infant mortality began to decline during the late 19th century. Although the cause probably lay in improved nutrition rather than in medical care, the fact that infant mortality was declining lent credence to the claims of doctors that the death of infants was preventable rather than a judgment from God. Further, during the early 20th century hospitals rather than homes were increasingly the scene of birth. In 1900 only about 5% of pregnant women delivered in hospitals. By 1939 half of all pregnant women and 75% of urban women delivered in hospitals. The transfer from home to hospital not only stimulated an increasingly number of medical interventions, such as cesarian sections, but also facilitated the registration of births. Registration was, in turn, indispensable to the realization of the idea that infant care could be made a science, for no one could prove that one approach was better than another until one knew how many infants were being born and what became of them.[10]

Relying on newly developed techniques of advertising, societies like the Child Health Organization broadcast the value of scientific care for pregnant mothers and infants. A coalition of physicians interested in public health and social workers took the lead in demanding federal intervention to establish childbirth and infant care on a scientific basis. The upshot was the Sheppard-Towner Act of 1921, which used the principle of matching grants to stimulate states to adopt the registration of births and a variety of principles of infant care. During the eight years of its existence, the Sheppard-Towner Act subsidized expanded birth and death registration, state divisions of child hygiene, and literally thousands of conferences in which pregnant mothers were told to restrict their diets and control their weight if they wanted to produce healthy babies.[11]

Even during the 1920s, however, the idea that scientific care could drastically alter infant mortality came under attack. The Sheppard-Towner Act had as stormy a history as any major piece of legislation relating to the American family. The most intense opposition to it emanated from the American

Medical Association, which took its cue from the fact that many of its advocates were social workers rather than physicians. What made matters worse was the fact that most of these social workers were women and, indeed, feminists organized around such characteristic institutions of the Progressive Era as the United States Children's Bureau. Viewing social workers as little better than socialists, the *Illinois Medical Journal* attacked the bill in 1921 as "another piece of destructive legislation sponsored by *endocrine perverts, derailed menopausics,* and a lot of other men and women who have been bitten by that fatal parasite, the *upliftus putrifaciens.*"[12]

It was not just the organized medical profession that attacked Sheppard-Towner. The congressional debates over the bill revealed an alliance between the medical profession and traditional conservatives who thought that mothers knew all that was necessary to give birth and care for newborns. They complained that the bill would substitute the schemes of reformers for all the sound experiences and customs of the past. These conservatives viewed Sheppard-Towner not merely as an attack on custom, but as an assault on privacy as well. Venting a number of overt and covert resentments, Senator James Reed of Missouri argued: "The poor, blushing little woman is compelled to register upon the public records the fact of her pregnancy. About that time, a bespectacled lady, nose sharpened by curiosity, official chin appointed and keen, is observed eagerly inspecting the ledger to ascertain the names of future mothers."[13] Whatever the good senator intended by this comment, it takes no search for subliminal meaning to detect a certain antifeminism behind it. His stereotype of the opposition reminds one of Southern stereotypes of Yankee school marms teaching freedmen during Reconstruction: in either case she was homely, single, indeed single at an age when she should have been married, and, by implication, channeling her sexual frustrations into a destructive busybodiness.

During the 1920s, in sum, advocates of the idea of federally sponsored scientific care for pregnant women and infants had to tiptoe through a minefield. Indeed the Sheppard-Towner Act probably would never have passed Congress had it not been for an historical accident. It was proposed one year after women were enfranchised to vote in federal elections and, although a majority of Congressmen probably opposed the bill in their hearts, they voted for it because they feared a backlash from female voters, a then unpredictable electoral force. Yet opposition, whether from the AMA, from anti-feminists, or from supporters of states rights, did not destroy the ideals behind Sheppard-Towner.[14] Even after the demise of the Act in 1929, advocates of scientific infant care continued to carry this case to the country through organizations like the National Congress of Parents and Teachers, the home economics divisions of the land-grant colleges, state divisions of child hygiene, and through periodicals like *Parents' Magazine*, which was launched in 1926 under subsidy by the Laura Spelman Rockefeller Memorial. Most of these institutions originated only after 1910, many only after 1920. Collectively, they comprised the motorized infantry units by which the Progressive ideal of applying science to the solution of social problems was disseminated widely throughout the nation.

The men and women who looked to science for solutions thought of science as a body of unarguable principles and unimpeachable facts, and they viewed their opponents as ignorant. As a corollary, they were confident that the diffusion of scientific knowledge would effect a radical transformation in the practice of infant care. In the view of advocates of science, child research stations like the University of Minnesota's Institute of Child Welfare would discover facts and principles; periodicals like *Parents' Magazine* would then tell everyone who would listen about these facts and principles. It was scarcely accidental that the official sponsors of *Parents' Magazine* included the child study centers at Yale, the University of Minnesota, the University of Iowa, and Teachers College.[15] Just as science was thought to be value free, unconditioned by cultural prejudice, so too the process of diffusing science was thought to be straightforward, a dissemination of unarguable facts that would sweep ignorance before them.

Whether this preoccupation with a science of infancy during the early 20th century actually produced a revolution in practice is hard to say. My guess is that it did not, simply because I find it difficult to believe that most parents could

follow consistently the exacting advice given them by specialists in infancy. Consider, for example, the message of *Infant
Care*, the famous pamphlet published in 1914 by the United
States Children's Bureau. Along with much wisdom about the
dangers of patent medicines (many of which were loaded
with opium) and the need to pasteurize milk, its author, Mrs.
Max West, passed along some remarkably ascetic ideas. Even
delicate babies, she said, should not be allowed to sleep in
rooms heated to more than 65°F. Between three and twelve
months, the baby should sleep in a room heated to only
fifty-five and in its second year in one heated to forty-five
degrees. Parents, in addition, had to discipline themselves
never to play with their babies or to use pacifiers. Toilet
training was to begin early, at three months, and in some
cases even earlier than three months. "In order to be effective," Mrs. West wrote, "the chamber [i.e. the pot] must be
presented to the baby at the same time every day, usually just
before the morning bath, and it must be presented per-
sistently each day until the habit is formed."[16]

We do not have to search for subliminal meanings to see
in Mrs. West some nascent behaviorism, nor to recognize in
her a residual Victorian moralism. Both Victorian moralism
and Watsonian behaviorism emphasized habit formation at
the earliest possible ages. Indulging the child was as big a
mistake as punishing it. The goal was the production of a kind
of automaton, the self-controlled, self-regulating republican
citizen. Almost by definition the ideas of a Mrs. West would
remain a minority opinion, for, as she readily conceded, they
assailed human nature, the normal parental response, at every turn. It was natural to want to pick up a crying infant;
parents had to learn not to pick it up.

None of this deterred Mrs. West or her successors from
their conviction that a correct science of infant care existed and
that this science would put infancy beyond the realm of controversy. There was no hint in *Infant Care* that rival ideas had
any legitimacy. The only alternative to science was ignorance.
Yet, as opposition to the Sheppard-Towner Act suggests, just
below the surface were resentments against reformers that
cannot be dismissed merely as offshoots of ignorance. Rather,
these judgments grew in part out of different cultural values,
values, for example, relating to the type of adult that children

were to become. To illustrate this point, let us begin by reminding ourselves that Mrs. West virtually equated mental
health with physical hygiene. Put differently, she assumed
that a correct physical regimen in infancy would produce a
child with healthy emotions. During the 1930s and 1940s,
however, it began to become clear that scientists themselves
were having difficulty in defining mental health or emotional
adjustment. Anthropologists like Ruth Benedict, Ralph Linton, and Margaret Mead demonstrated that definitions of the
good person or the healthy child were culturally conditioned.
Some cultures emphasized orderliness, others spontaneity.
Some aggressiveness, others acquiescence. Some achievement by the individual, others conformity to the group. A
scientist might prefer one type of personality to another, but
science itself did not possess a single best system to communicate to parents and inevitably some scientists began to
tell parents to relax and follow their own instincts.[17]

Conflicts about the appropriate role of women in society
also contributed to controversies over infant and child care.
Not only have conflicts relating to feminism had a long history, but that history has repeatedly intersected the history of
infancy. Both in England and New England, for example,
judges and juries were much more likely to convict a woman
for infanticide if she was "wanton" (i.e., single) and the child a
bastard.[18] Similarly, the movement between 1840 and 1880 to
stiffen hitherto lax laws against abortion in America coincided
with and reflected hostility to feminism. As James C. Mohr
has noted in his study of abortion in 19th-century America,
toward the middle of the 19th century abortion increasingly
became a device of family limitation among middle-class married women rather than merely a recourse of unmarried girls
desperate to avoid social ostracism. As the clientele of abortionists changed, the medical profession often blamed the rise
in abortion on the spread of feminist notions that had ren-
dered middle-class women too self-absorbed to accept the
sacrifices of motherhood. This presupposed, of course, that
such women were getting abortions furtively, without knowledge of their husbands. It is unlikely that this presupposition
had much basis in reality; as Mohr notes, there is no record of
husbands suing wives for getting secret abortions. But mere
absence of evidence did not deter mid-Victorian physicians

from writing as if feminist propaganda had polluted the
minds of American wives and plunged them into the alleys in
search of abortionists and in flight from their divinely
ordained maternal role.[19]

In summary, despite the longstanding search for a science of infancy, differing ethical and cultural values have
disrupted efforts to vanquish controversy from the field. The
conflict over abortion during the nineteenth century was one
sign of the way in which science itself has been laden with
values. At this juncture, allow me to try to set the abortion
issue in a somewhat broader context. Abortion is one among a
number of ways that have emerged over the course of time to
deal with unwanted pregnancy. Alternative modes have included infanticide, abandonment, and putting a child up for
adoption. It may seem jarring to link all of these together, for
abortion today is more widely tolerated than either infanticide
or abandonment. But, putting aside moral judgments, they
are all means to the same end. What strikes the historian is the
extent to which the acceptance of any one of these methods
has been culturally conditioned. The acceptance of each, in
other words, has fluctuated. In England, for example, penal-
ties against infanticide became much more severe during the
sixteenth and seventeenth centuries, and then loosened both
in England and New England during the late eighteenth century. As noted earlier, laws against abortion became much
more severe in the late nineteenth century. Interestingly,
during the late nineteenth century, at the very time when
abortion was becoming a less acceptable way of dealing with
unwanted pregnancy, adoption was rising in estimation.
Voluntary associations like the Women's Christian Temper-
ance Union began to sponsor homes for unwed mothers,
often called "Anchorages," that oversaw adoptions after the
girl had delivered the baby.[20]

What accounted for the rise of Anchorages? Part of the
answer certainly lies in the existence of a growing number of
middle-class parents who desired to adopt children and hence
created a market for the services of Anchorages. The presence
of this market cannot be separated, in turn, from changing
conceptions of the function of children in middle-class families during the late nineteenth century. Children were no

longer viewed primarily as economic assets but as stimulants to the unlocking of the affections, to warmth and family feeling. And for this purpose nothing was better suited than an infant, economically useless in the extreme, but emotionally attractive by virtue of its very helplessness. As Viviana Zelizer has noted, at the very time when the child was becoming worthless, he was also becoming priceless.[21] The existence of a growing number of clients, married couples eager to adopt infants, undergirded the efforts of Anchorages to make adoption a preferred way to deal with unwanted pregnancies. Of course the practice of adoption did not spring into existence for the first time during the late nineteenth century. There had been adoptions during the colonial period. But before the late nineteenth century most adopting parents appear to have been related to the adopted child, and the latter was usually not an infant but between the ages of three and ten. After 1850, in contrast, the adopted child was likely to be an infant unrelated to the adopting parents.[22]

Not only have differing ethical and cultural values led to controversy over the care of infants, in sum, but they have also accounted for changes in the acceptability of varying ways of dealing with unwanted pregnancy. From one perspective, the history of infancy has been the history of the search for a science of infancy that would substitute reason for ignorance. Those who have engaged in this search believed that, rightly understood, science would overcome infant mortality, ensure healthy infant care, and guarantee happier motherhood. Advocates of scientific infant care during the early twentieth century not only put their trust in disseminating pamphlets on proper infant care, but also thought that science would come close to solving the problem of unwanted pregnancy. Organizations of the Progressive Era like the American Society for Sanitary and Moral Prophylaxis promoted programs of sex education in the public schools with the aim of reducing sexual activity among adolescents (and hence illegitimacy) by disseminating correct information about sexuality. Science would, in effect, inoculate against desire. From a different perspective, however, the history of infancy has been a good deal messier than the scientists of infancy hoped. Efforts to change practices relating to infancy in a scientific direction have encountered resistance spawned by shifting and often conflicting cultural and ethical norms. The best that can be said for what I have called the science of infancy is that it has been one among a number of views competing for popular acceptance.

Inasmuch as controversy over infancy is longstanding, one might conclude that controversies today merely revive traditional controversies. In some ways, this generalization is accurate. Abortion, for example, was certainly a heated issue in 1850 or 1880. Although physicians since the eighteenth century have called for a science of parturition and infancy that would put these processes beyond the court of public opinion, public opinion has had a nasty way of intruding itself on the well-laid plans of the medical community. Yet to note that the abortion issue has some significant historical antecedents is not quite the same as saying that controversies today merely rehearse past controversies, for there are some major differences between the debate over abortion now and the debate a century ago. The most notable of these is probably the fact that feminists have switched sides. Throughout the second half of the 19th century, feminists often found themselves allied with antifeminist physicians in opposition to abortion. Even allowing for differences today within the feminist movement, there has been a monumental shift on the subject.[23]

Why did 19th-century feminists line up against abortion? The key lies, I think, in their utopian view of equal marriage, what one historian has called their commitment to "true love and perfect union." An article of faith among 19th-century feminists was their belief that virtually all of the unhappiness in marriage resulted from the social and economic inequality of the partners.[24] The more radical the feminist, the more intense was this belief. When men and women were equal in all ways, they argued, men would finally accord "respect" to women. Here I have to play the role of translator: "Respect" meant that men would become less sensual. They would learn to control their appetites. As soon as they did, divorce, birth control, and abortion would disappear. As a corollary, 19th-century feminists viewed these customs as byproducts of the corrupt status quo. Abortion, in particular, was a repulsive practice to feminists, an example of the degrading conduct that inequality forced on women. It might be necessary, but it was a necessary evil. Its existence testified to the sensuality of men that was born of dominance and that would disappear as soon as dominance by men was put to flight.

It goes without saying that these views no longer describe the position of feminists. To trace the reasons for the shift would carry us far afield into the history of feminism in America. For the moment, let me content myself with a modest observation. The shift in the attitude of feminists has coincided with an intensification of controversies surrounding public policy toward the family. As the debate over the Sheppard-Towner Act in the 1920s makes clear, controversies are not new. But they occur now on a much grander scale. If we can cut through Senator Reed's layers of rhetoric, we can see that he did not like the idea of educated, single women telling the federal government what to do, and he did not want the federal government to tell states or private citizens what to do. In sum, he did not like intrusiveness. The position he represented is still around. In 1976 the Republican Party platform took a swipe at the *Roe* decision by complaining about the intrusiveness of federal judges in our lives. But that complaint today is merely a drop in the sea of controversy that engulfs public policy toward the family.

To take one illustration, in 1909 the first White House Conference on child welfare assembled. Periodic White House conferences during the next two decades provided Progressive social workers, sociologists, reformers, and public-health doctors with a convenient forum for addressing the country on the need for rational, scientific solutions to problems like juvenile delinquency and infant mortality. One of the pastimes of these conferences was drawing up lists of goals in the field of child welfare, which were grandiloquently termed children's charters or bills of rights for children. Many people continued to view these reformers as visionaries, and perhaps for that reason no one spent much time attacking them. The reformers themselves agreed on most issues. The White House conferences were like coronation processions for liberal ideas about the family. From time to time, of course, reformers miscalculated and tripped one of the mines in the field of public opinion. In 1924, for example, they convinced Congress to draft a constitutional amendment banning child labor. To the surprise of most reformers, this amendment touched off a storm of opposition, mainly from the Roman Catholic Church, and it was buried under an avalanche in state legislature after state legislature. But even during the 1920s, opposition to reformers were spasmodic and unpredictable. To mix a metaphor, the land mines had not yet become lobbies.

In contrast, the last two decades have witnessed a remarkable politicization of public opinion on issues relating to the family. If you don't believe this, compare for a moment the consensus that marked the White House conferences on the family in 1909 and 1930 with the turmoil that surrounded President Carter's plans for such a conference. The only suitable analogy I can think of to convey the sense of anger that suffused Carter's conference is to compare it to the early meetings of the United Nations at Lake Placid, when Soviet and American delegates would wrangle for weeks not about specific policies but about the agenda. In fact, the Carter conference began to unravel over a similar issue. Not only did delegates have difficulty agreeing on an agenda, but they could not agree on a definition of the family.

To summarize, the history of childbirth and infancy in America reveals two central and conflicting trends. The first has been a movement to put all issues relating to infancy on a scientific and presumably unarguable basis. The architects of this position assumed that once scientific laws had been established and diffused, customs based on ignorance and prejudice would melt away. Science, in their view, was a source not merely of authority but of uniformity. That uniformity, however, has proven elusive. In addition to the trend toward a scientific outlook, there has been a growing trend toward controversy. Once controversies flared over resentment of the intrusiveness of the self-conscious advocates of a scientific outlook. In recent decades, however, several new types of controversy have been added to this traditional type. Public policy toward the family is now as politicized as policies on the tariff were a century ago. In this context, the construction of a science of infancy seems not only elusive but illusory as well.

A Note on Monsters

One sign of the growing quest to extend human control over the events surrounding birth was the gradual erosion of the traditional doctrine that many defective newborns were "monsters," half-human, half-beast, and without a right to life. In his *Commentaries on the Laws of England* (Book II, Section IV), William Blackstone described "a monster which hath not the shape of mankind, but in any part evidently bears the resemblance of a brute creation, hath no inheritable blood, and cannot be heir to any land, albeit brought forth in marriage." As late as 1854 an American medical journal reported that lawyers commonly distinguished between monsters and deformed newborns with human heads and a right to life.[25] By then, however, the distinction was starting to collapse. It was probably not accidental that the same journal attributed the distinction to lawyers, while itself defining a monster as merely an extremely defective newborn. By the mid-nineteenth century, works on medical jurisprudence were converging on the doctrine that "individuals are not allowed to destroy these monstrous births."[26] Yet authorities on medical jurisprudence conceded that among the vulgar the idea still prevailed that it was not illegal to destroy monstrous infants, and the authorities themselves contended that such destruction would usually be manslaughter rather than murder.

Notes and References

[1] C. John Sommerville, *The Rise and Fall of Childhood* (Beverly Hills, California, 1982) passim.

[2] Ibid.

[3] Ibid.

[4] Edward Shorter, *The Making of the Modern Family* (New York, 1975), pp. 168–204; George D. Sussman, "The Wet Nursing Business in 19th-Century France," *French Historical Studies, 9* (April, 1975), 305–322.

[5] Richard W. Wertz and Dorothy C. Wertz, *Lying-In: A History of Childbirth in America* (New York, 1977), p. 22.

[6] Lawrence Stone, *The Family, Sex and Marriage in England, 1500–1800* (New York, 1977), pp. 464–466.

[7]Wertz and Wertz, *Lying-In.*

[8]Jacqueline S. Reinier, "Rearing the Republican Child: Attitudes and Practices in Post-Revolutionary Philadelphia," *William and Mary Quarterly* 39 (Jan., 1982), 160.

[9]Wertz and Wertz, *Lying-In.*

[10]Ibid., p. 133.

[11]S. Josephine Baker, "The First Year of the Sheppard-Towner Act," *Survey,* 52 (April 15, 1924), 89–91.

[12]Quoted in Robert Bremner, *et al., Children and Youth in America: A Documentary History* (Cambridge, 1971), II, part 7, 1020.

[13]Quoted in ibid., p. 1016

[14]J. Stanley Lemons, *The Woman Citizen: Social Feminism in the 1920s* (Urbana, 1973), chap. 6.

[15]Orville G. Brim, *Education for Child Rearing* (New York, 1959), pp. 328–337.

[16]Mrs. Max West, *Infant Care* (U.S. Department of Labor, Children's Bureau, Care of Children Series #2, Washington, D.C., 1914), p. 51.

[17]Ruth Benedict, *Patterns of Culture* (New York, 1934); Ralph Linton, *The Cultural Background of Personality* (New York, 1945).

[18]Peter C. Hoffer and N. E. H. Hull, *Murdering Mothers: Infanticide in England and New England, 1558–1803* (New York, 1981) chaps. 1, 2.

[19]James C. Mohr, Abortion in America: *The Origins and Evolution of National Policy, 1800–1900* (New York, 1978) chap. 4.

[20]Joan Brumberg, "'Ruined': Family and Community Responses to Adolescent Out-of Wedlock Pregnancy in Upstate New York, 1890–1907." Paper presented at the Annual Meeting of the Organization of American Historians, April 6, 1983.

[21]Viviana Zelizer, "The Price and Value of Children: The Case of Children's Insurance," *American Journal of Sociology,* 86 (1981), 1037.

[22]Jamil Zainaldin, "Child Exchange in Boston: The Origins of Modern Adoption, 1851–1900." Ph.D. thesis, History, University of Chicago, 1976.

[23]Mohr, *Abortion in America,* pp. 111–13

[24]William Leach, *True Love and Perfect Union: The 19th-Century Feminist Critique of Marriage* (New York; Basic Books, 1980).

[25]"Upon the Legal and Social Rights of Malformed Beings," *American Journal of the Medical Sciences,* 28 (July, 1854), 276–277. (I am indebted to Professor James C. Mohr of the University of Maryland, Baltimore County, for calling my attention to this reference.)

[26]Alfred S. Taylor, *Medical Jurisprudence* (4th American edition, Philadelphia, 1856), pp. 357–358.

Response to "Science and Controversy in the History of Infancy in America"

Margaret O'Brien Steinfels

What intellectuals or journalists, or for that matter physicians or psychologists, report about child-rearing is not necessarily an accurate description of what parents do to their children, or how they raise them, think of them, or feel about them. Professor Kett's opening, cautionary remark underlines a problem that historians have in turning to the past when trying to answer questions about the relationship between parents and children; about the details of child care; about the necessary distinctions to be made among child neglect, child abuse, and infanticide. Not only is it difficult to recover the details of domestic life (the fashion of using wet nurses, for example) or the meaning of public events (the establishment of foundling hospitals), it is nearly impossible to uncover people's motives in using them. Institutions or services founded for one purpose are often used by clients for other or additional reasons. Foundling hospitals were established to care for abandoned newborns and they became the repositories for illegitimate and orphaned children as well. What were the motives of mothers in leaving their children at the foundling door? Surely not that those mothers believed the hospitals to be charnel houses, but rather that they believed the infant or child had a greater chance of survival in the care of these hospitals.

Another example: the day care centers established in the last years of the 19th century in New York City and other American cities had many high-minded purposes: to Americanize the children of immigrants, to provide proper hygiene, to teach good manners and orderly behavior. Parents probably did not object to those purposes, but they often used the centers for practical purposes: the centers were better than leaving children home alone; the centers could more easily wash and feed infants and children than could parents living in old-law tenements without water and sometimes without heat. Although the centers diapproved of mothers working for any reason other than dire need, many women went to work to improve the family's standard of living.[1]

If we are to be sensitive to the difficulties in uncovering motives and meanings, as we must be in drawing comparisons between the past and present, our judgments ought to be temperated by the recognition that most parents then, and most parents now, were and are endowed with a spontaneous desire to protect and nurture their young; in that I do not think we are remarkably different from animals. Improved conditions of life, nutrition, medical care, and so on, simply make it easier nowadays for parents and others who care for the young to actually protect them in more effective ways than they could in centuries past. That infant mortality has declined, that children are not maimed or deformed by disease or malnutrition, that they do not suffer from so many accidents is not necessarily a sign that the 20th century has produced better parenting and child-care, but primarily I believe a sign of material and technical advance.

The air of fatalism that Professor Kett mentions—a fatalism that permeated parental attitudes in past centuries—was real enough and often expressed, for example, in the diaries and other writings of 17th and 18th century men and women. Ann Bradstreet's poems mourn her possible death in childbirth, allude to the dangers of illness and accident in childhood, and grieve for the death of her daughter-in-law and grandchild in childbirth. Michael Wigglesworth's diary entry for May 16, 1656 recounts the prolonged, but successful, labor of his wife in birthing a daughter in words of pity, sorrow, joy, and humble thanks for their survival.[2]

But why should this air of fatalism necessarily lead to indifference? The fact that death, accident, and disease could strike at any time, with few resources for evading or alleviating them, might make parents cautious and careful. Samuel Sewall records throughout his diary a solicitous and hovering attitude toward his wife, his children, his servants, and a genuine grief over the death, at birth or shortly thereafter, of several infants who are mourned by the community and properly entombed. John Demos in a survey of early New England court records finds no evidence that child abuse or neglect even existed. In a concluding remark, he notes: "Had individual children suffered severe abuse at the hands of their parents in early New England, other adults would have been disposed to respond. The culture, in general, seems to have sponsored a solicitious attitude toward the young. . . ."[3]

Sending infants to wet nurses was not uncommon among middle- and upper-class families in England and Europe; it is often cited as an example of the indifferent regard in which parents held newborns. But look at the case of Samuel Johnson's mother, Sarah Johnson. She gave birth to Sam, her first-born, at the age of forty; he was not a vigorous newborn and Michael Johnson insisted that because of Sarah's poor milk supply the baby be sent to a wet nurse, Joan Marklew. She lived a few hundred yards from the Johnson house. There Sam was visited at least daily by his mother who, fearing the neighbors would laugh at her concern, devised various routes to reach the Marklew house. To Sam's weak constitution his wet nurse added scrofula, which left him partially blind. Throughout his childhood, his parents sought the ministrations of medical men for his condition, even going so far as to travel to London in hopes that the Queen's touch would cure Sam's disease, also known as the King's evil. Some of these parental decisions may have been mistaken, but it is hard to see them as the reflections of an indifferent attitude toward a sickly infant; in fact, given the difficulties of travel and the vagaries of medical advice, at the time, the Johnson's efforts could be described as heroic.[4]

These examples suggest that indifference does not flow from a lack of control over the health and well-being of one's children; fatalism is not necessarily followed by neglect. Lack of control and the everyday experience of sudden death may have conditioned parents to greater care rather than less.

These examples also suggest that in all human activities culture is a crucial mediator of our spontaneous or instinctual behaviors. The instinct to protect the newborn and the young has always been a central focus of parental care. What varies over time are the cultural and psychological pressures that support or attenuate those "natural" behaviors. I think this offers a fair interpretation of Sarah Johnson's behavior. She bowed to the cultural pressures of the time by sending Sam to a wet nurse; she was convinced yet anxious; she came to think that Sam's chances of survival were greater with Joan Marklew's more abundant supply of milk. She did what many modern parents do: faced with conflicting courses of action, she acquiesced in the one that promised most benefit to Sam, even though it was one that made her very unhappy.

What many parents believe to be best for baby often turns on that kind of social pressure; it is what their culture, their physician, their radio talk show hosts say is best for baby. Swaddling was once *de rigueur*. Not so long ago, obstetricians restricted weigh gain during pregnancy to twenty pounds. It now appears that this was mistaken, that weight gain must vary on an individual basis, too little being perhaps more dangerous than too much. Anesthetized childbirth was once thought to be good for mothers, convenient for physicians and nurses, and benign for babies. I believe that most physicians now recognize that general anesthesia during delivery can be detrimental to a baby. Who knows? Perhaps the present fashion of cesarian deliveries will come to seem excessive or unnecessary. My examples could be multiplied. The point is this: that except for the generally healthier condition of mothers and infants today, some modern medical practices could also be described as dangerous or as reflecting indifference, much as many practices common to past centuries now seem to us.

Today it is true that the health and well-being of most infants has been assured through a higher standard of living, improved nutrition, and better medical care. Nonetheless, I agree with Professor Kett that a science of infant care is not only elusive, it is illusory. That such a science can provide guidelines for our decisions, ethical and otherwise, seems to be more illusory the more deeply medicine moves into technological fixes for treating ever smaller and premature newborns, and of caring for genetically and congenitally impaired infants.

In fact, it sometimes seems that, in its present spurt of advances in genetic and neonatal medicine, medicine sets before us a contradictory patchwork of choices and standards. For example, selective abortion for genetic disorders is now available and supported by many physicians. Through amniocentesis and chromosomal screening, physicians can identify fetuses with certain genetic or metabolic disorders. If the parents so choose, physicians will abort these fetuses in the latter part of the second trimester and sometimes at the beginning of the third trimester, that is, at 15 to 24 weeks gestation. But what happens if an infant is born with a defect that might have been detected through amniocentesis, such as Down's syndrome or spina bifida? What if medical or surgical intervention is needed to preserve or enhance the life of this genetically impaired infant? The recent cases of Baby Doe in Indiana and Baby Jane in Long Island are relevant examples. The temptation not to provide medical care to such an infant is great: it could have been aborted 12 to 21 weeks before its birth; our society does not provide adequate social and material support to children who are handicapped; the breakdown of small communities and kinship networks make it likely that the parents alone must face the task of caring for and finding their child lovable, often against the stigmatizing pressures of neighbors, schools, and so on. In the absence of compelling moral values, an ever-diminishing number of physicians and surgeons will be able to help parents consider the value of their child in the face of serious physical or mental impairment.

Professor Kett reminds us that abortion is only one of a number of ways that unwanted fetuses and unwanted pregnancies have been dealt with through history. Infanticide, abandonment, and adoption, too, have served this purpose. He also reminds us how culturally conditioned the ascendancy of one or another of these methods is in different periods of history. For future historians, the question may be: Why was abortion the ascendent method in the last decades of the 20th century?

Abortion today is medically safe and it is legal; and for most people it is a more acceptable solution to inopportune pregnancies than infanticide. But as I have tried to suggest, the link between selective abortions and selective infanticide is present in the logic of amniocentesis and genetic screening. Absent legal sanctions, I do not see how selective abortion for genetic defects will not lead to selective infanticide for the same reasons.

In conclusion, let me return to an earlier point: our attitudes toward children, toward child-rearing, toward becoming parents are culturally mediated. As individuals, we may choose or not to become parents, but our society, our social world, our culture all signal that some children are not only more desirable than others, but that some children are so undesirable that we can find no reason to support their existence.

Parents standing in a nursery over a newborn infant with a serious disability know that, for the most part, our society considers that child undesirable. So too, do adolescent mothers receive such a message; and often enough, parents who are poor receive the same message. Children do not become unwanted and undesirable simply because their parents feel that way, but because by our inclination to stigmatize, by our welfare policies, and health care policy, our political and economic policies (our unemployment policies, for example), we as a society signal to some parents that their children are undesirable and unwanted.

We have difficulty in knowing how true that was of the seventeenth century; the high infant mortality rate, the primitive understanding of medicine, the likelihood that only the most hardy survived in any event may make it a moot point. But it is true today. Our hubris is in thinking that because we know more we will somehow do better.

Notes and References

[1]Margaret O. Steinfels, *Who's Minding the Children? The History and Politics of Day Care in America.* New York; Simon & Schuster, 1974, Chapters 2 and 3.

[2]Donald M. Scott and Bernard Wishy, ed., *America's Families: A Documentary History.* New York: Harper & Row, 1982, p. 139, and p. 130.

[3]John Demos, "Child Abuse in Context: An Historian's Perspective." Unpublished paper, January 1979. Quoted with author's permission.

[4]John Wain, *Samuel Johnson: A Biography.* New York: The Viking Press, 1974, pp. 17–26.

Comment on "Science and Controversy in the History of Infancy in America"

Alan J. Weisbard

I write here not as a representative of a humanistic discipline as usually defined, but as a lawyer whose scholarly work and prior experience with the staff of the President's Commission on Ethics in Medicine and Research have focused, in part, on the legal and public policy dimensions of the care of imperiled newborns. What I hope to do in these comments is to make somewhat more explicit several possible implications of Professor Kett's historical findings for contemporary discussion of social policy and legal interventions in this area. I will try to satisfy the commentator's obligation to be at least a bit provocative in the process.

Differences in Values

One major thrust of Professor Kett's analysis is the historical trend "to put all issues relating to infancy on a scientific and presumably unarguable basis"—resting on the assumption "that once scientific laws had been established and diffused, customs based on ignorance and prejudice would melt away." As Dr. Fleischman's article makes clear, enormous strides have been made in our technical capacities to care for imperiled newborns. But the triumphs of scientific knowledge and medical technique have by no means led to any uniformity of moral values. Indeed, as we find ourselves making hard choices in the neonatal intensive care wards—

47

choices previously made for us by nature and which, as Professor Kett reminds us, were viewed historically as part of God's plan, outside the sphere of human control—we are constantly reminded of the pluralism of values in this society, even on matters so fundamental as the life or death of an infant.

Interestingly enough, much of the impetus toward recognizing differences in values has come from within the medical–scientific community, perhaps symbolized most powerfully by the publication of Duff and Campbell's[1] pathbreaking article on neonatal care in the *New England Journal of Medicine* in 1973, and by their continued strong advocacy of decision-making by parents in conjunction with physicians, without the intervention of courts or other legal institutions. This recognition of competing values, and willingness to work with, and ultimately accede to, the values of the parents, has been endorsed by the Judicial Council of the AMA[2] and is further evidenced by social science polling data of physician-respondents.[3]

This data, incidentally, shows a particular inclination by many members of the medical profession (at least prior to the recent controversy) to respect parental decisions *not* to treat physical defects of infants subject to mental retardation, such as Down's Syndrome children. It is, perhaps, an interesting commentary on our times that less than 10 years ago, at the time of the Karen Quinlan decision, doctors were being castigated for refusing to give up, for favoring prolongation of life over all considerations of quality of life as experienced or valued by the patient. Today—as evidenced by the Report issued by the President's Commission in March of 1983[4]—it is apparently felt necessary to encourage physicians to accept the desire of *parents* to hope against hope—to pursue vigorous therapy even when it is likely to prove futile—while recognizing that some physicians "find it personally offensive to engage" in such treatment and may seek to withdraw from such cases. A recent Massachusetts survey finds that a majority of pediatricians would not *recommend* surgery for a Down's infant with an intestinal blockage.[5] Whatever one thinks about that position—and the Commission explicitly rejected it in the Down's case—values are apparently changing, both within the medical profession and in the wider society.

Differences in Means

A second theme, and for me perhaps the single most striking observation in Professor Kett's discussion, is the historical linkage among several alternative modes of dealing with unwanted pregnancies and children: abortion, infanticide, abandonment, and adoption. All of these, as Professor Kett points out, may be viewed as "means to the same end"—at least if one puts aside moral judgments. To be sure, in our own time, moral judgments of these various alternatives *do* differ. But it is precisely here that the historical dimension provides a challenging perspective on contemporary debate by reminding us that popular and/or societal acceptance of these methods has fluctuated over time, and that attitudes toward them are culturally conditioned.

One question arising from this observation is the *relevance* of the fact that some of our attitudes toward legal and moral questions regarding the care of imperiled newborns *are* culturally conditioned. These historical data provide something of a response to the contemporary argument that failing to provide life-sustaining care to imperiled newborns can only be understood as yet another element of the moral decay of our times—that in the past, before the challenges to religious authority and traditional social norms that are characteristic of our times, such things could not have happened. As Professor Kett has indicated, such things *did* happen in the bad old days of Victorian England and Puritan America.

But should our present-day moral repugnance toward, for example, infanticide, be overcome by the realization that this repugnance is, at least in part, culturally conditioned? I would argue not. While recognizing the evils of past cultural imperialism, I think we can and should stop well short of any thoroughgoing moral relativism. Most of us do care—and properly so—about the values that characterize our own culture, and rightly believe that we can and shall be judged by the

values we proclaim and defend. I am prepared to state my
own preference for, and commitment to, a society that places
priority on defending the interests of the most vulnerable—
those who cannot protect themselves. The *hard* problem is
recognizing what those interests are and who should decide
—questions raised directly by Dr. Fleischman.

How are the four alternatives noted by Professor Kett to
be evaluated in our present context? Clearly, the question
invites fuller discussion than is possible here. I shall thus
content myself with a few remarks designed to provoke further discussion and debate that relate the present topic to the
still-raging controversy over abortion.

Contrary to much popular opinion, the Supreme Court
did not rest its recognition of the abortion right established in
Roe v. Wade[6] on a woman's "right to control her own body."
Indeed, the Court expressly noted that "it is not clear to us that
the claim . . . that one has an unlimited right to do with one's
body as one pleases bears a close relationship to the right of
privacy previously articulated in the Court's decisions."[7]
Rather, the Court founded its application of the emerging
constitutional right of privacy to decisions whether or not to
terminate a pregnancy on the following catalog of interests:

> Specific and direct harm medically diagnosable even in early
> pregnancy may be involved. Maternity, or additional offspring,
> may force upon the woman a distressful life and future. Psycho-
> logical harm may be imminent. Mental and physical health may be
> taxed by child care. There is also the distress, for all concerned,
> associated with the unwanted child, and there is the problem of
> bringing a child into a family already unable, psychologically and
> otherwise, to care for it. In other cases, as in this one, the addition-
> al difficulties and continuing stigma of unwed motherhood may be
> involved. All these are factors the woman and her responsible
> physician necessarily will consider in consultation.[8]

It would certainly appear that many of the maternal or parental interests considered by the Court to justify abortions could
be carried over to support parental decisions to decline medical treatment for imperiled infants. Given the Court's rejection of distinctions based on the woman's control over her

body, the basis for rejecting a parental right to decline treatment, or indeed, a parental right to infanticide, must rest on the moral and legal status of the infant as distinct from that of the fetus. A number of philosophic commentators have challenged the moral basis for any such distinction, contending that the difference is merely one of "geography," not of moral status; some have gone on to argue that infants and young children should not be regarded as persons in any full moral or legal sense.[9] I suspect that the most satisfactory responses to such arguments rest not on philosophical logic, but on psychological and cultural understandings of the meaning of birth and admission to the human community. The historical perspective provided by Professor Kett is relevant here, and I hope it will stimulate further exploration of the deeper, cultural significance of birth and, perhaps, of the somewhat mysterious conception of "fetal viability" promulgated by the Supreme Court in *Roe v. Wade*.

Recognizing that much more can be said on abortion and infanticide, let me move on to the final two of Professor Kett's four alternatives: adoption and abandonment. Adoption has been a strangely neglected alternative in many recent discussions of both abortion and the withholding of life-saving care from imperiled newborns. The Supreme Court has had rather little to say about the possibility of adoptive placements as alternatives to abortions—perhaps because the enormous number of abortions performed in this country raises serious questions about the availability of sufficient adoptive placements. But an interesting aspect of several celebrated recent cases of withholding treatment from newborns has been the lining up of families eager to adopt—as in the Bloomington Baby Doe case. Perhaps for the reasons elucidated in *Roe v. Wade*, the biological parents should have the legal ability to avoid continuing responsibility for the care and support of an unwanted child—a legally permitted form of "abandonment," perhaps justifiable in terms of the interests of both parents and infant. But who, or what, has conferred upon such parents the right to choose death for imperiled newborns, when physicians are willing and able to sustain life and other families are eager to adopt?

Current Government Policy

With these points in mind, I would like to turn to certain
aspects of current Administration policy toward care of imperiled newborns. As is well known, the Reagan Administration's response to the distressing case of Bloomington Baby
Doe in Indiana was the promulgation by the Department of
Health and Human Services of a series of regulations pur-
suant to §504 of the Rehabilitation Act of 1973. These regulations are said to be designed to protect handicapped infants
whose lives may be endangered by discriminatory denials of
food or customary medical care. The regulations have already
been extensively criticized as crudely conceived, deeply
ambiguous in meaning and prospective application, and failing to make necessary distinctions among the different classes
of cases to which the regulations are apparently addressed.
Many of these criticisms seem to me just and well-taken. Let
me elaborate briefly:

If one believes, as I do, that decisions regarding the care
of imperiled newborns are of different kinds—that some such
decisions, although painful and agonizing to the parties, are
morally straightforward—here I would include, on the one
hand, decisions in favor of surgically treating correctible
physical defects of Down's children, and on the other hand,
providing comfort but *not* making futile and inhumane efforts
to prolong the life (and suffering) of a doomed anencephalic
infant—while other situations have no clear moral resolution—and here I would include many of the decisions about
when to continue, and when to stop, treatment of extremely
low birth weight babies with multiple anomalies and a highly
questionable prognosis—then the approach of the HHS regulations appears grotesquely oversimplified and inadequate to
morally sensible decision-making. Given these differences, it
is a bit difficult to see what these regulations achieve other
than a highly public display of the Reagan Administration's
solicitude for sick babies. This public display bespeaks a
rather selective concern for handicapped newborns by an
Administration that has made unprecedented cutbacks in our
commitments to supporting the vulnerable and dis-
advantaged in our society. Some have said that this Adminis-

tration's humanitarian concerns begin at conception and terminate at birth. I suppose these regulations put the lie to that proposition by demonstrating that the Administration's humanitarian concerns now extend beyond birth to discharge from the neonatal ICU. That, however, is apparently the end of the line.

Though, as I have said, these criticisms of the HHS regulations appear well taken, some others seem, to me at least, to go too far. Thus, in the unlikely event there is anyone left in the audience whom I have not yet provoked, let me offer a few more favorable words on the regulations that, although not a defense, may constitute a lawyer's plea in mitigation.

First, some of the other criticisms that have been advanced seem to me misguided or excessive. For example, the *New York Times*, in an otherwise excellent recent editorial whose conclusions were largely identical to my own and to those of the President's Commission, characterized the regulations as "in effect set[ting] up a spy system in the nursery."[10] Although the rhetoric was dramatic, I wonder whether the *Times* would similarly characterize the child abuse and neglect reporting laws already in effect in every state in this country, many of which already potentially apply to nontreatment decisions in the neonatal nursery. To be sure, the HHS regulations injected an adversary tone that all involved would have preferred to avoid. And alternative mechanisms *are* available to forestall unnecessary or premature intrusions by the State into the decision-making process. I will address those alternatives later in this article. However, we should not forget that the newborn cannot protect its own interests. If parents and physicians reach a decision seriously adverse to the interests of the child, which society is potentially prepared to overturn, there must be some vehicle to trigger outside intervention on behalf of the child—some way for information to reach persons in a position to protect the infant. Overdramatic rhetoric about "spy systems" does little to resolve the difficult questions of when and how to trigger such independent inquiry.

It has also been stated that these regulations appear anomalous in intruding on personal liberty when propounded by an Administration dedicated to "getting government off people's backs." Were this a regulation limiting the rights of competent adults to make decisions regarding their own medical treatment, that charge would clearly have merit. But it is by no means clear to me that the objection makes sense when applied to an effort to protect the interests of babies unable to protect themselves, and whose parents may have powerfully conflicting interests of their own. Though the current HHS regulations may do a poor job in achieving their stated purposes, there is a legitimate role for law to play in protecting the vulnerable.

So much for excessive criticism. What can be said in favor of the Administration's regulatory efforts? Just this: In their very failure to do the job, these regulations have brought serious public attention to the issue and have challenged the rest of us to find better and more effective ways to protect the vulnerable who cannot protect themselves. Let me suggest three mechanisms worthy of consideration, which parallel those recommended by the President's Commission.

Mechanisms for Protecting the Vulnerable

First, the creation of mechanisms to enhance the ability of parents to participate knowledgeably and effectively in decision-making regarding the care of their child. This goes far beyond obtaining their signatures on an informed consent form. It incorporates the creation and training in empathetic, social, and communication skills of a team of neonatal professionals to provide both information and emotional support to parents suddenly thrust into what may be the most difficult decisions of their lives. Since the need for decisions may be sudden and may involve issues with which the parents are unfamiliar, it may be especially important to help parents to understand the opportunities and possibilities open to handicapped children and their families, the availability of community resources, and, where possible, to have the opportunity to meet with other parents who have lived through similar decisions. Further, information should be provided on the possibilities of adoptive placements for such children.

Second, the establishment of explicit hospital policies and processes to assure careful and considered decision-making regarding the care of imperiled newborns. If hospitals do not do a satisfactory job of tackling these issues, others will take them over. The threat of bedside decision-making by those who conceived the recent regulations should be a potent inducement to action by hospitals and other social institutions to abort any further such conceptions and find a better way. Creative experimentation with independent hospital ethics committees and cooperative collaboration with child protective agencies have great potential in this process.

Third, a commitment to improve the human and financial resources we devote to imperiled newborns after they leave the neonatal ICU—and to the families, natural or adoptive, that will care for them. If we wish, as a society, to enshrine the preciousness of life and the centrality of human dignity among our values—as I think we ought—then we must provide the support and meet the special-care requirements necessary to assure lives of dignity, opportunity, and meaning to the precious infants we save and to the families who love and care for them.

References

[1]Raymond S. Duff and A. G. M. Campbell, "Moral and Ethical Dilemmas in the Special Care Nursery," 289 *New Engl. J. Med.* 890 (1973).

[2]Judicial Council of the American Medical Association, *Current Opinions,* section 2.10, Chicago (1982).

[3]See, e.g., "Treating the Defective Newborn: A Survey of Physicians' Attitudes," 6 *Hastings Center Report* 2 (April 1976); Anthony Shaw et al., "Ethical Issues in Pediatric Surgery: A National Survey of Pediatricians and Pediatric Surgeons," 60 *Pediatrics* 588 (Supp. 1977); David Todres et al., "Pediatricians' Attitudes Affecting Decision-Making in Defective Newborns," 60 *Pediatrics* 197 (1977).

[4]President's Commission for the Study of Ethical Problems in Medicine and Biomedical and Behavioral Research, *Deciding to Forego Life-Sustaining Treatment: A Report on the Ethical, Medical and Legal Issues in Treatment Decisions,* Washington, DC (1983). See especially Chapter 6: "Seriously Ill Newborns," pp. 197–229.

[5]See Todres et al., *supra* note 3.

[6]410 US 113 (1973).

[7]Id. p. 154.

[8]Id. p. 153.

[9]See especially the work of philosophers Peter Singer and Michael Tooley.

[10]*New York Times* (April 2, 1983), p. 18.

Section II

Religion, Suffering, and Morality

Our Religious Traditions and the Treatment of Infants

David H. Smith

Introduction

The babies we are talking about have problems, and cause problems. They can only make the heart ache, and commentary written by outsiders can easily chafe the wounds. This may be especially true when the commentary is said to reflect the religious sensibilities or traditions of our people, for religion means to be about life and death and truth. We sometimes are outraged by religious claims and arguments, but usually that is because our expectations for religious insight and principle are so high. We chastize the goddess who has failed us. I understand my role to be to offer one ordering of religious perspectives on the care of impaired infants. Because the issues are so powerful and so complex, I shall strike at them in several different ways. I hope the result has a kind of unity or continuity, but I think it is more likely to be the coherence of a symphony rather than the logical tightness of a Euclidean proof. Indeed, I shall play several of the same notes that Professor Arras has struck in his presentation. But they are important notes, and my chord structure is rather different.

My general plan in the remarks that follow goes something like this. I shall begin by discussing some of the images or portraits of children and parents that rest in the consciousness of Western religious people. These portraits, I think, have informed our thinking about human relationships in profound ways. They provide some of the data from which theological moralists should begin their reflection. Of course, like all portraiture, they reflect the perspective of the painter; these are not value-neutral descriptions. Moreover, the "morals" of the incidents I am about to mention are not always clear and unambiguous. This ambiguity may well be part of their greatness, but it is also a limitation for purposes of moral analysis. The portraits are not a sufficient moral framework, nor even a sufficient basis for such thinking. Still, they are illuminating and constitutive of the life of our central traditions.

With these data spread out before us, I shall presume to extract some general principles from them. And I will apply these principles to some of the moral issues that have already been raised so eloquently for us.

Jewish and Christian Foundations

At age 40 Isaac married Rebekah; their sons Esau and Jacob were born soon thereafter. "When the boys grew up, Esau was a skillful hunter, a man of the field, while Jacob was a quiet man, dwelling in tents. Isaac loved Esau, because he ate of his game; but Rebekah loved Jacob." (Gen. 25:27f) Children here, and elsewhere in the Bible, have a kind of particularity and parents respond to them with partiality. They like and dislike, pick favorites. This human tendency, the stuff of great literature and soap opera, is treated as an inevitable fact in the scriptural narratives.

In fact throughout our history the individuality of a child is related to the child's lineage. This is not always fortunate for the child. In the book of Judges, Jepthah swears to kill the first person he sees on returning home if only the Lord will help him to triumph over the Ammonites. When he does come home, victorious, "behold his daughter came out to meet him with timbrels and with dances; she was his only child; beside her he had neither son nor daughter." With her agreement he sacrifices her "for I have opened my mouth to the Lord and I cannot take back my vow." (Judges 11:29f) We shall return to the question of parental vows in a moment; for the moment I note that the fate of Jepthah's daughter is appropriately—if unhappily—determined by the behavior of her parent. Children are often victims in the Biblical narratives and images (cf. Psalms 137:9—"Happy shall be he who takes your [i.e., Babylonian] little ones and dashes them against the rock"). They are victims of a particularity, or peculiarity, over which they have no control. This is a not altogether happy or benevolent aspect of reality, in the Biblical portraits.

This stress on particularity has not always been reflected in Western consciousness about children. In his rich study, *Centuries of Childhood*, Phillippe Aries notes that hundreds of years passed before Western Europeans really discovered the special characteristics of childhood. Before the 17th century, he writes:

> *No one thought of keeping a picture of a child if that child had either lived to grow to adulthood or had died in infancy. In the first case, childhood was simply an unimportant phase of which there was no need to keep any record; in the second case, that of the dead child, it was thought that the little thing which had disappeared so soon in life was not worthy of remembrance: there were far too many children whose survival was problematical. . . . one had several children in order to keep just a few. . . . People could not allow themselves to become too attached to something that was regarded as a probable loss. (p. 38)*

Increasingly, Aries observes, people began to realize that childhood was a special stage of life, valuable in its own right. This led to treating children as pets, to coddling, but increasingly to a notion that children had to be educated or disciplined so as to grow up into happy adults. Ultimately, however, only increased medical success as well as contraception have been able in the past 200 years to bring about a time in our society in which children are valued in and for themselves. Really taking particularity seriously was a long time coming.

This mention of education gets at another religious image of the child—the child as one whose future must be prepared for and whose character and education therefore matter. Beginning with the embryonic passover traditions in Exodus 12 and continuing right through the passover aggadah of the present time, Jewish life has always wanted to help the child situate himself as part of a people with a destiny, to have a sense of himself as part of a community with a past, present, and future. The great 20th century Protestant theologian Karl Barth expresses this idea very well. Children, he says "are not by nature their (parents') property, subjects, servants or even pupils, but their apprentices, who are entrusted and sub-ordinated to them in order that they might lead them into the way of life." (III/4 p. 243) Jewish or Christian parenthood inevitably looks to the child's future with a vision of the good life.

Thirdly, and most centrally for our purposes, the religious images of childhood and parenting that we know are—as the quotation I just read from Barth suggests—images of *limited* parental authority and dominion. When the baby Moses was three months old his mother put him in a basket and let him be taken over by others for his well being (Exodus 2). At age twelve Jesus in the temple is not a very good boy, straying from his parents and telling them that he really has more important things to do than go home with the family. Mary and Joseph's authority over him is distinctly limited—indeed, if he is *the* paradigmatic child he vividly illustrates the ambivalence of parent/child relations for he both is, and is not, *their* child.

This limitation of parental authority is nowhere more vivid than in the story of the binding (akedah) of Isaac by Abraham in Genesis 22. We are used to thinking of this story in terms of Isaac's loss of life, but of course it would have been Abraham's sacrifice in more ways than one. He is being asked to give up that which means the very most to him; it is a story of letting go. Abraham looks up and sees the ram and God explains "now I know that you fear God, seeing you have not withheld your son, your only son, from me." (Genesis 22) Isaac's life is not, ultimately, Abraham's to dispose of as he may see fit, for Abraham would never have chosen this sacri-fice. As a planning parent, Abraham serves a master, and his parental power is limited.

This brings us to a fourth image of childhood that we find in our heritages. Elijah the Tishbite stays with a widow whose son dies. Elijah "stretched himself upon the child three times" and the child revived. (1 Kings 17) The earliest history of

Christianity tells the story of a boy named Eutyches (Lucky) who fell asleep in a third story window while listening to a long sermon. He falls out of the window and dies, but Paul (the preacher) revives him with an embrace. (Acts 20:8–12) Children are presented as having access to Jesus and his contact with them in at least one version of the story is wonderfully physical and bodily: he takes them in his arms and blesses them. (Mark 10–16 and parallels) It is not the child's personality alone, nor his future, but his body that is the focus of concern in these images. And, for Christians, this image is intensified even more in the extraordinary claim for an identification of God with the body of a crying, soiled human baby. The *baby* Jesus is one of the two or three dominant Christian portraits.

Children in these images—and I have not pretended to give an encyclopedic account—are embodied human beings with a particular past and future. They create responsibilities for their parents. These parents, however, are not given complete dominion over their children, only a "slight seniority." (Barth III/4 p. 246)

The Morals of the Stories

These images do not of themselves lead us to moral conclusions. They only inform us; they require an ordering. I propose to use the notion of fidelity as an ordering principle. I shall assume that the great theme of Western religion is God's loyalty to us—a loyalty responded to by betrayal and disobedience, occasionally by trust and love. A life of passionate devotion to God should involve compassionate loyalty to other persons: this is the central theme of Western religious morality. What does a faithful working out of the images I have sketched entail?

When we try to figure out how to be faithful to people, one thing these images push us to think about is the things that they live for, what—as we say—makes them tick. Thus when it comes time to shop for presents for a birthday or Christmas or Chanukha we look for presents that relate to the interests of a person. A set of records of Beethoven symphonies are marvelous for one person, a waste for someone else. Someone likes jewelry, someone else a set of golf clubs. People's interests or, as I would prefer to say, their loyalties, vary. Loyalty to someone means respecting this kind of particularity. And the respect may go quite far; my love for my friend may well bring me to love what he loves, to care for those things that matter to him.

Usually we know something of a person's specificity from his or her explicitly stated preferences. We know that Dad likes to play golf because he says he does, and he says he does not care for classical music. In fact, however, this correlation between what people really are and who or what they say they are is not one-to-one. Many people *say* they like to do things that in fact it is clear that they do not much enjoy; anyone who has ever lived in a family can write a book about the various kinds of self-deception that human beings indulge in. Still, throughout most of our lives, we are willing to acknowledge that a person's decisions about his or her own life ought, at least, to have a preferred place. Medical decisions are no exception. Thus we support a right to refuse treatment and stress the importance of a requirement of informed consent.

The striking thing about babies, however, is that part of this stress on moral particularity cannot exist for them. They have never had a chance to care about or live for anything. They have not established a style of life or character. It is not just that because of great defect some babies will never have the kind of personality that a normal person would have. It's rather that they have never got off the launching pad. The effect on our moral reasoning about them is that an important variable simply drops out of the picture. We see them as children whose individuality has not yet emerged.

They also begin with a minimal medical history. Their prognosis is at best uncertain. We do not know what degree of retardation a child with Down's syndrome will have, or how serious the deformity and hydrocephalus associated with myelomeningocele will turn out to be. Some premature babies have terrible and short lives; others do not. It is never possible to be absolutely sure of the outcome.

I draw from these facts the conclusion that for the most part we owe it to these babies to get them started. They are cursed by powers in their lineage over which no one has control. But the kinds of personal style and preference arguments that might justify a decision for the death of an adult can never be applied directly to newborns, and it takes time to establish a kind of pattern of biological functioning such that we can extrapolate to future events and responses with some fairly good degree of plausibility.

A second thing that loyalty to another person means is interest in or concern for what he or she may become. Hence the power of the image of parent as educator. The idea perhaps is difficult. We tend to form *set* pictures of people; "Oh, he could do it if he wanted to, but he will never skip Monday night football." We think of people as set in their ways: *He* is punctual, *she* is stubborn, *they* are argumentative. It is true that the characters that people have possess a kind of constancy, but it is also true that people change and grow. Saints fall away, sinners convert, Prince Hal becomes Henry IV.

When we look into the future for our friends we find that we want many things for them. One of these is *happiness*. This is a notoriously hard term to define, and I am not capable of producing a statement that will instantly persuade everyone. But I think it obvious that happiness is something that we want for those whom we care about. We would describe as *sick* someone who wanted his child or spouse or parents to be *un*happy. Naturally I do not mean that we want happiness of any kind for them—we may well feel that some kinds of happiness are better than others. Certainly our religious traditions suggest the importance of piety before God as an ingredient in true happiness.

A second thing that we want for our friends is *excellence*. We hope that they will do something well. This something may be athletic, intellectual, or social. Some are good runners, others good thinkers, others are, as we say, good people. In whatever way we define it, however, excellence is something that we want for our friends, and for religious persons it is associated with some kind of relation to God.

Given these goals, the question becomes whether they

are attainable. Seldom do we know the answer to this. This is particularly true with handicapped babies. We may be certain that they will *not* have the kinds of happiness or excellence that is open to normal children, *but this is scarcely the issue.* We are not trying to compare them with others for not all children of God play the same role in the kingdom. Who among us would survive some great assize trying to decide if his own life, with all its defects, was worth living? The issue is, is some kind of happiness or excellence open to this child? As the custodians of their future we may act with hope. Our role is to help them know themselves and God, whatever their future may hold.

Related to this, I should mention that this stress on hope and the future is especially important when families are stretched and torn as they are in the cases we have been discussing. On the one hand, the parents need to stick with the child; on the other hand, the medical staff needs to stick with the family. Care is not an episodic or momentary thing. To be genuine it must involve a commitment to what is to come. Followup is very important. The uncertain character of the future is frightening, especially to parents whose dreams and expectations have been smashed. Loyal medical care for families in this situation involves living through this uncertainty with them and, through one's presence, giving them hope.

Third, loyal care for persons involves care for their *bodies*. Children in the religious images are not just personalities, nor bundles of potentiality; they are living, struggling bodies. There are two aspects of this obvious point to which I wish to call attention. One of them is that the body has a kind of life of its own. It is not altogether malleable to human desire, whether we speak of the desire of its "owner" or of someone else. And health is comparably objective.

Let me put the issue a little differently. Bodies are not just things we have, they are things we *are.* The requirements of a healthy body can be generalized—they are not completely relative to cultural prejudices or values. And medical care involves care for bodies in a special way. Not to the exclusion of everything else. Not in some scientific way that excludes the caress and the cuddle—but oriented to the body all the same. My lawyer is a consultant on my rights, my tax advisor on my money, and my doctor on my body and its health. Medical care for a baby means making him or her comfortable.

Secondly, an important feature of bodies is that they change. Bodies pass through stages, regardless of the desires of anyone. These include infancy, adolescence, middle age, aging, and dying. Proper forms of care should be adjusted to these various stages or moments in a life span. Pediatrics is not just medicine on small adults; we rightly use different terms to describe a highway accident to children, other adults, and the aged. Fidelity to a patient will always require optimal care, but as we change, so do the requirements of care. Throughout most of our lives medical loyalty to us means keeping us alive, but this ceases to be true at some point in our lives. Then we may actually speak of a need to die. Dr. Anna Fletcher has referred to a baby who was "trying to die."

Paul Ramsey has made this point with characteristic force. He says that the determination of when a person enters dying is a "medical" decision.* This means that there are changes in the ill person himself that determine what the right forms of treatment are. When a person starts to die, then our responsibilities shift. Ramsey's idea, which is rooted in religious images, is that a person's life trajectory at some point enters a dying phase. When that happens moral responsibility changes. Then the issue is, "Is he dying?" or—more generally—"What forms of treatment are optimal care for her in whatever time she has left?"

If the discernment of when someone begins to die—a determination that is relevant to the choice of appropriate forms of care—is a matter of what Eric Cassell has called the "healer's art," the substance of the point is best captured not

---

*Sometimes Ramsey suggests that it is a strictly biological matter. This is mistaken. While there are clinical signs that one is dying (as there are of the onset of puberty), the actual discernment of the beginning of the process is largely a matter of art. Experience and skill are relevant to the determination, of course. Many of us (including some physicians) are tone deaf to what is happening in the body of another. But there is no escaping the factor of judgment, of discerning perception. People's judgments will differ, to be sure, but this fact in no way proves the non-existence of dying as a last stage of life.

in argument but in metaphor. A couple invite friends to dinner. Food and drink are pleasant; the conversation bubbles. The good host is hospitable and courteous to his guest, no matter what his shifts in mood. But there comes a time when the party "winds down"—a time to acknowledge that the evening is over. At that point, not easily determined by clock, conversation, or basal metabolism, the good host does not press his guest to stay, but lets him go. Indeed he may have *to signal* that it is acceptable to leave. A good host will never be sure of his timing and will never kick out his guest. His jurisdiction over the guest is limited to taking care and permitting departure.

Analogously, loyalty to other persons involves care for them of the best possible sort, and changing the forms of care as needs change. It means recognizing that a time comes when we can care for them no longer, only bid them godspeed. We may, as it were, show them the door. But we should not kill a patient, because that would be to betray him, to assume the kind of jurisdiction over his fate that is incompatible with being a good host. And we can never be sure of our timing.

Loyalty to the dying, in other words, is compatible with a choice of palliative clinical care and personal human support over life-extending technologies. It is of the greatest possible importance that this shift be seen as a shift to an alternative *medical* form of care and it is the great strength of the hospice movement to have refined and institutionalized these alternative forms of medical care. Loyalty to the dying not only tolerates but positively mandates this shift; making a guest stay longer than is good for him is very bad manners.

Anyone attending this conference realizes the possibility that some children are born dying. This may be simply in virtue of prematurity. Thus I think we should simply comfort some newborns as they die. On the other hand, I have meant to suggest that any predictable defect, in particular retardation, is not a sufficient reason for saying a child is dying. It is especially important not to yield to pressure to come to a quick and efficient decision in order to spare people's feelings. It takes time to discern what is going on with a young and small patient's body.

This brings us to the last point I wish to make. *Loyalty to a defective baby requires involvement in a decision-making procedure of integrity and credibility.*

Some writers on medical ethics see this as the only issue. In the case of adults they stress the patient's right to die; for children they assert an absolute right of parents to make the decision. However, it is clear that in the Biblical images the gift of a child does not entitle parents to unlimited sovereignty over it. Abraham has to let go, despite his plans and hopes. Even if parents are the child's best proxy in virtue of their identification with it, their power should not be unchecked. There are other reasons for this. Parental judgment is finite and it is possible that the parent's identification and stake in the issue will produce excessive bias. Further the communal nature of human existence implies that we do not live by ourselves; the family is not an island. Desertion of the family—and simply leaving the decision in their hands may amount to desertion—is a form of betrayal.

In effect I am saying that loyalty in these situations requires patience. There is no rushing the decision-making procedure, and no substitute for involvement of the physician throughout it. What's needed is lots of talking between parents, physicians, nurses, and appropriately concerned others.

This is no panacea. Even a good process can go wrong or be misused. I can well understand—indeed I largely share—the impulse to establish minimal standards in the law. For in this area discretion abused is not just indiscreet; it is immoral. I do not think babies should be let go simply in virtue of Down's syndrome—as I hope I have suggested. But at the moment I cannot imagine a regulatory net or law that would ensure a discriminate result or could be fine tuned to all the variety of cases we want to acknowledge. Thus I fall back on moral education and a process of listening, learning, and support.

In sum: loyalty is what the peoples of God owe a defective baby. We owe him respect and hope, care and comfort for his body, fair play, and due process. Sometimes this will mean we have to kiss him goodbye. But never without having made him welcome, never without a hug, and never without regret.

Acknowledgments

Portions of this paper were first presented at a conference I attended in Nebraska in November 1982. I am grateful to all the participants in that meeting and especially to Dr. Robert Grant, Director of Maternal and Child Health for the Nebraska State Health Department, for the occasion to listen and think together.

"Suffer the Little Children . . ."

Suffering and Neonatal Intensive Care

Thomas H. Murray

Introduction

An Irish priest was once describing with great dis-approval a practice common among his fellow priests and brothers who were also teachers in the early grades, the practice of caning or striking the child with a thin flexible rod. He remarked that these men, having read that passage in the New Testament that says "Suffer the little children to come unto me . . ." had taken it literally. It is the suffering of the littlest of the little children, not by caning but by accidents of conception and birth, and by the actions of their medical caretakers, that is my topic. Not merely the suffering of the children themselves, but also the suffering caused to their parents and family, and to the medical staff into whose care they are entrusted. Along the way I want to discuss some features of suffering in infants that are not unique to, but perhaps are still acutely true of, this group of sufferers.

First, let me sketch very briefly the dimensions of neona-tal intensive care: the numbers of infant-patients, doctors, and dollars devoted to it, as well as its impact. Roughly 6% of live-born children go to intensive care—200,000 babies a year. For many, the only need is for a brief course of therapy for jaundice, or extra nutrition to put on weight and gain strength. For others, though, the stay is longer, and the re-sults often uncertain. The average length of stay is just under two weeks, although stays of over a year are recorded. There are approximately 600 hospitals with neonatal infant care units (NICUs), 7500 beds, and 800 neonatologists,[1] which a recent report concludes is 200 fewer than currently needed.[2] Total patient days are estimated at 2.6 million, and the yearly cost of that care at around $1.5 billion (that's $8000 per patient). Leaving aside the brief and inexpensive stays (those costing less than $4000 in one year), neonatal cases were the most costly of all high-cost hospitalizations recently studied, averaging over $20,000. This was considerably higher than the care of neoplastic diseases, and comparable to end-stage renal disease and coronary bypass surgery.[3]

What does this enormous investment of effort and money buy? In 1915, one child in 10 died in the first year of life. By 1950, the comparable number was three in 100; by 1979, it had been further reduced to slightly more than one in 100. How much of this reduction occurred in the first month of life? From one in 20 in 1915, to one in 50 in 1950, to one in 115 by 1979.[4] Of this reduction, just how much results from neonatal intensive care and how much from other factors, is difficult to say. From 1915 to 1950, the bulk of the drop in infant mortality was the result of better care of those children from one month to one year old. Since 1950, there has been an equally significant improvement in saving babies in the first month. Indeed, the improvement in the neonatal mortality rate may be even better than it appears from the numbers I have just cited. Before the advent of sophisticated neonatal care, many babies who today would be resuscitated, only to die shortly thereafter and appear in the mortality statistics, would have been classified as "stillborn," and not counted as a live birth at all. Neonatal intensive care cannot claim credit for all the improvement in infant mortality, but it seems indisputable that today, babies can be saved who could not have been just a few years ago, especially those born very prematurely and severely underweight.

Suffering of the Babies

Against the background of the magnitude and impact of neonatal intensive care, we can now ask the central questions about suffering. There are three groups whose suffering concerns us here; the babies themselves, their parents and families, and the hospital staff who care for them. To begin with those just beginning, do the babies suffer, and if so, what forms does that suffering take?

Hospitals, whatever else they are, are also places of suffering. People come there *because* they are suffering, in hope of relief. When they get there, they often discover new forms of suffering in the medical procedures themselves, in their loss of privacy and autonomy, in their isolation from those whom they love and on whom they depend, in the loss of familiar objects and the daily routines that are often sources of comfort.

Of the sorts of illness- and hospital-related suffering I have just mentioned, all but one—the suffering inflicted by painful medical procedures—do not seem to apply to newborns because they assume a personal history, a memory of things being other than they now are, a hope that they may be different, or even the capacity to imagine them different. All these are cognitive abilities that we do not assume newborn infants have, especially those who appear in NICUs, whose neurological capacities may be damaged or less mature than healthy newborns'.

Yet, it would be wrong to conclude that our infant-patients in the NICU do not suffer, or at least do not experience pain. I have heard some—not all—neonatologists minimize the newborn's sensitivity to pain, but I believe this is an understandable reaction to the realities of their daily work—probing, inserting, testing. Curiously, one of the leading texts in Neonatology—Avery's—contains virtually no references to the infant's pain, and none at all to suffering.[5] What discussion there is of pain concerns the mother's pain in childbirth, the drugs used to relieve it, and the impact of those drugs on the baby. But T. Berry Brazelton has this to say: "Compared to any other species, the human neonate is relatively helpless in the motor sphere, and relatively complex, even precocious, in the sensory sphere."[6] Is it reasonable to think that a sensitivity to pain might be part of that "precocious sensory sphere?" In addition, we have the evidence of our own observations. Newborns react to pain; they flinch,

their face contorts, they cry. Even the comatose infant may
have some capacity to experience pain. One neonatologist
with whom I was discussing this question remarked that he
would never consider operating on a child, even one whose
brain had been devastated, without using anesthesia. The
conclusion seems to be that infants do experience pain, or at
least we should always treat them as if they could.

The law has a mixed record in dealing with this issue. A
1938 case, *Babb v. Murray*, had refused to allow damages for
pain and suffering caused to a 3-½ month infant on the
grounds that the child could not know what was happening,
and hence could not experience fear or anguish.[7] Over thirty
years later, the same court (California Supreme Court) had
this to say about a related case: "The infant's inability to
explain the cause of pain or to describe the extent of it does not
affect the sting of it. Indeed, the infant's cry of pain is as
poignant as the most detailed exposition. The moan of the
injured child, who may even be unconscious, needs no
elaboration in descriptive language. Communication flows
from all manner of sounds and gestures; it is not confined to
brittle and inadequate words. The inarticulate anguish of the
infant serves as much as a ground for recovery as the adult's
most sophisticated description."[8] In California at least, infants
not only feel pain, they also suffer.

There are important distinctions between the two, which
are pertinent to the NICU, but which I can discuss only
briefly. One could argue that although infants experience
pain, suffering requires something in addition—a history that
has known the absence of pain, a language for comprehend-
ing and expressing the fact that I have pain and that it might be
otherwise, an expectation of pain continuing into the future.
Eric Cassell, a physician, seems to suggest as much when he
writes, "Suffering occurs when an impending destruction of
the person is perceived; it continues until the threat of disintegration has passed or until the integrity of the person can
be restored . . ."[9] It seems unlikely that a newborn's consciousness could bear the weight of perceiving "a threat of
disintegration" or its "impending destruction." David H.
Smith, a theologian, is more explicit: "For suffering to occur,
there must be a subject who can comprehend or interpret his

experience . . . In other words, we suffer because we have a perspective on the world that can be threatened by events."[10] Again, the newborn's mental equipment is just not up to the task.

The problem is this: In wanting to show that suffering is not synonymous with physical pain, but is much more deeply linked with our core as persons, Cassell and Smith are led to deny the simplest root meaning of suffering. As the *Oxford English Dictionary* defines it, suffering is "The bearing or undergoing of pain, distress, or tribulation," or simply "A painful condition."[11] I think Cassell and Smith are right in saying that suffering can be much more than mere physical pain—indeed many kinds of pain when they are chosen, or accepted as a part of the pursuit of a valuable goal, are not perceived as suffering, and may be dismissed as "mere pain." After all, it is the pain that is not understood, that seems random, and for which no end can be envisioned that produces the greatest suffering. It is, in other words, pain as an infant experiences it.

My conclusion is that infants do indeed suffer, or at least should be treated as if they did, and I agree with Cassell and Smith that relieving suffering is one of the fundamental goals of medicine, and that it is useful to think of "the practice of medicine as a form of ministry to the suffering."[12] The fact that newborns may not be so successful in registering their objections to unrelieved suffering should not make their suffering any less our concern, or its alleviation any less important to us.

The Parents' Suffering

Parents of the babies who enter NICUs also suffer. They may suffer even if the baby does not! A friend who runs a NICU told me of a pattern that perplexed him. He claims that parents of babies who are taken into the NICU for hyperbilirubinemia—jaundice to the layperson—are the most anxious of all, and appear to suffer intensely. Uncomplicated anemia in newborns is about as nonthreatening a problem as anyone could wish. A little time under the bili-light, the color improves, and the baby goes home. Why are the parents so troubled? My guess is that their anxiety is connected to the same expectations that lead most people to think not at all about the possibility that their child might be born imperfect. We joke with the couple when they announce the pregnancy, tease the woman about her altering appearance, prepare a place for the baby, celebrate the birth, and bring mother and infant home triumphant. There is no room in this scenario for delay or imperfection. When they arise, they are painful blows.

Parents of more seriously ill infants suffer in many more ways, and I can mention only a few. Their general anxiety, fear, and concern for the welfare of their child will be aggravated by their limited ability to understand the medical information about their child, and especially their uncertainty about the future health or capacities of their baby.

Very likely they will be worrying about their own role in causing the child's problem: Is it a defect in one of the parent's genes? Did the mother smoke or drink too much? Take a drug she should not have taken? Would it have made a difference if they had watched her diet more carefully? Been more scrupulous about seeing the obstetrician and following his or her advice? Seen a better obstetrician?

They may well be angry at the baby for being imperfect, for causing them all this distress. They may be angry also at the medical staff for coming between them and their child, for hurting their child, for keeping it alive. Anger like this may arouse guilt for feeling angry.

The parents' lives will be dislocated; they may have to alter their schedules to accommodate visits to the hospital and slight their jobs, or other children, or pursuits that they prize; or they will have to decide not to do so, not to be with their sick infant, and accept the guilt that flows from that decision.

Perhaps the most painful early suffering is that accompanying the decision whether or not to terminate treatment for their baby or, if the baby lives, to accept that child into the home or turn it over to the care of the state or an institution. For all the attention these decisions receive, they are relatively rare, and a very small share of the parents of NICU babies are ever seriously faced with them. Still, decisions like these determine whether or not we will create a large pool of chronic sufferers in the future. Handicaps likely to be created include severe retardation, cerebral palsy, seizure disorders, and blindness. A few years ago, for babies weighing 1500 grams or less, about one-fourth of the survivors suffered some serious handicap, but this proportion is decreasing. It translates into approximately 9000 such babies a year. This is a large part, though not all, of the chronic medical problems borne by infants passing through NICUs.

So far, I have mentioned sources, forms, and indicia of suffering likely to occur in the first days of encounter with the NICU. Other elements appear only later. There is an abundance of evidence that bringing home a severely damaged or defective child can be very stressful to a family. Marriages break up; mothers, fathers, and siblings suffer a variety of emotional problems. These studies are often countered by anecdotes about couples who swear that their marriage has been strengthened and their lives deepened by their experience. This undoubtedly occurs; the point we should take is not that these exemplary couples have not suffered—they certainly have—but that they have somehow transcended their suffering, or to put it more neutrally, they have coped with it differently from others. Trauma certainly occurs, but we probably know less about its forms, intensity, and prevalence than we might think.

One further note about the prospect of taking a baby home: In many ways, it is much easier on the family if the baby dies than if it comes home seriously disabled, or is sent to live in some institution. I doubt that any parent confronts the idea of caring for a severely damaged child with unalloyed joy, and that person would be not quite human if he or she did not fleetingly entertain the thought that life would be much easier without the child. A parent may come to feel that he or she has wished the child's death; the potential sense of guilt from this should be obvious.

One profoundly tragic situation that I can do no more than mention occurs when a late abortion results in a live-born fetus that is then sent to the NICU. Relations between parents and staff can be very tense. This leads me to the last group, the medical staff.

The NICU Staff and Suffering

I think it is necessary to talk separately about the nurses and the doctors. Some observers note the high turnover rate of nursing staff on NICUs, and attribute this to the stress caused by working on such units in close proximity to suffering parents and babies. I am not so sure. NICUs tend to be staffed by young nurses, freshly graduated, technically sophisticated, and desirous of living in the large metropolitan centers where NICUs are usually located. Whether or not the special stresses of working in NICUs contribute to the alleged high turnover rate is difficult to say. What is clear is that watching babies die is not easy; I suspect it is harder, perhaps much harder, than watching old people die, or even those in their middle years. My guess is that nurses rapidly learn to distinguish the winners from the losers, and concentrate their emotional energy on the survivors. I recall one infant who was doing well, whose parents visited infrequently, and who had been "adopted" in a way by the nurses, to the point where they bought and dressed him in the tiniest jogging outfit I have ever seen.

As one nurse explained to me, though, any time a nurse becomes attached to an infant, she or he "loses" it eventually, either because the infant dies or is discharged. Though the nurse is likely to be very happy when her charge goes home to capable and loving parents, there are painful cases where the parents appear to be unable to care for or cope with their infant. Joy Hinson Penticuff describes the dilemma: "Nurses worry about these little ones, in whose lives they have been so invested over days or months. Yet NICU nurses are not able to change these families, and they learn through the 'grapevine' that Johnny was admitted through the emergency room last night with skull and rib fractures—a victim of child abuse."[13]

Difficult as it is to find descriptions of nurses suffering on NICUs, it is even more difficult to find descriptions of physicians' suffering. However, an article by a psychiatrist attached to a large NICU offers some insights into the stresses experienced by neonatologists and neonatology residents. In response to a questionnaire, the unit's veteran neonatologists rated worries about inconsistencies in treatment approaches and feelings that they did not have enough time to talk with parents as sources of stress. House officers were more troubled by having to talk with parents—especially having to discuss the infant's illness or death—and having to do the neonatologist's job without adequate knowledge or experience.[14]

Physicians, having many more patients in their care, and unable to spend much time with any one, may not form as many, or as intense, attachments as nurses do. On the other hand, physicians are more likely to be the ones who must talk with parents about unfavorable prognoses for their child, or inform them their child will die or has died. Physicians bear the burden of ordering a course of treatment—or nontreatment—for the infant, and must accept responsibility if that decision is in error, medically or morally. Nurses may find themselves ordered to execute a course of treatment with which they may disagree—medically or morally.

With all that, there is a special, suffering-related problem posed to nurses and doctors on NICUs: how to form the emotional calluses that permit one to continue to function in spite of the frequent death of babies all around you, without losing the capacity to suffer yourself at these deaths, without taking the calluses home with you, creating an emotional distance from your own family, and without losing your ability to empathize with the parents of your charges. The suffering worries me, but so does the possibility that one might cease to suffer, and the human price that would extract from self, family, the babies, and the babies' parents.

Suffering and Morality

There are two final places where suffering touches our moral life. For one thing, in all the argument over when, if ever, treatment should be terminated for an infant, one of the generally agreed-upon justifications relies on a judgment that the infant's own suffering outweighs any benefit to be derived from the treatment. That is, the infant's suffering is a significant moral element in decisions whether to treat or not. We are not ethically obligated to prolong a life suffused by pain

and suffering. Indeed, we may well be obligated not to stand
in death's way in such cases.

John Arras's paper in this volume analyzes the role suffering plays as a moral justification for nontreatment decisions, so I will not discuss it further.

The Problem of Theodicy

The last thing I want to discuss is framed between the
comments of two anonymous parents of dying children. The
first: "So I go to hell for not believing. I've got news for you.
I'm already there."[15] The second happens to be by a minister.
"How happy we are that Rad is going to heaven soon . . . You
don't know what it's like to be a man of God, happy that God
has chosen your child for an angel."[16]

The problem even has a technical name, coined by Leib-
nitz: Theodicy—literally, "God's justice." It is the defense of
God in the face of the evil existing in the world. Kant described
it as "the defense of the supreme wisdom of the author of the
world against the accusations brought against it by reason on
the ground of the repugnant in the world." Hume put the
dilemma succinctly; he argued that it cannot be true both that
evil exists and that God is all powerful and perfectly good. No
form of evil threatens the contemporary belief in a God of this
kind, or for that matter, a belief on whatever grounds in the
fundamental goodness or justice of the world, more than the
suffering of children. The best St. Augustine could do to
explain the suffering of innocents was to say that it purified
and tested them, and proved humankind's solidarity in original sin.[17]

Father Paneloux, in Albert Camus' novel *The Plague,* was
a Jesuit priest and not coincidentally a scholar of St. Augustine's work. Paneloux delivers two major sermons to the
people of the plague-ridden city of Oran, the first after only
one month of the plague, the second many months later, just
before the plague has begun to loosen its hold. He opens his
first sermon this way: "Calamity has come on you, my brethren, and, my brethren, you deserved it." At one point he says
confidently: "The just man need have no fear, but the evildoer
has good cause to tremble." Paneloux later joined one of the

sanitation squads, who risk plague themselves in order to transport victims and enforce the quarantine. A turning point came for him when he witnessed the death of a small child, M. Othon's boy. It is a particularly horrible death, the boy frail though he is battling the plague ferociously, and suffering terribly. Paneloux's second sermon has a "gentler, more thoughtful tone . . ." He uses the pronoun "we" rather than "you." And nothing, he said, was "more important on earth than a child's suffering, the horror it inspires in us, and the reasons we must find to account for it." "The agony of a child," he said, was "humiliating to the heart and to the mind."[18]

Some deaths are more unjust than others. Some suffering seems more outrageous, more a blow to the coherence of the universe. To parents of a newborn who expected the event to be a source of joy for themselves and for their baby, to now have to witness their child's suffering, helplessly, is as severe a challenge to their faith in an ordered universe as I, at least, can imagine, whether one believes that the order is by God's hand or otherwise. And I think this too is a form of suffering, for which I have no name, but I believe deserves our serious attention.

I have tried to make five points.

First, the babies themselves probably suffer; in any case, they should be treated as if they did. Relieving their suffering is an important goal of medicine and should be a major concern in neonatal medicine.

Second, parents' suffering can be very deep and complex—a potent mixture of anger, grief, and guilt. They may be destroyed or ennobled by it, but there is no denying that for most parents, a dead, damaged, or endangered newborn is a severe blow.

Third, it is important for medical caretakers to maintain that tension between becoming used to the suffering and death around them on the one hand, and remaining open to being moved by it, on the other.

Fourth, suffering is a weighty element in the moral equation leading to the decision of whether to terminate or deny treatment.

Fifth and last, the suffering of innocents challenges our

conception of an orderly, just universe—or at least as many of
us who still cling to a belief in a just universe.

Notes and References

[1]Peter Budetti, Peggy McManus, Nancy Barrand, and Lu Ann Heinen, *The Costs and Effectiveness of Neonatal Intensive Care*, Case Study #10 of Background Paper, "Case Studies of Medical Technologies" for Congress of the US Office of Technology Assessment's study, "The Implication of Cost-Effectiveness Analysis of Medical Technology," Washington DC, US Government Printing Office, 1981.

[2]Committee on Fetus and Newborn, Committee on the Section on Perinatal Pediatrics, "Estimates of Need and Recommendations for Personnel in Neonatal Pediatrics," *Pediatrics*, 65 (4), April 1980, 850–853.

[3]S. A. Schroeder et al., "Frequency and Clinical Descriptions of High-Cost Patients in 17 Acute-Care Hospitals," *New England Journal of Medicine*, 300, 1979, 1306–1309.

[4]Budetti et al., op. cit.

[5]Gordon B. Avery, ed. *Neonatology*, 2nd edition, Lippincott, Philadelphia, 1981.

[6]T. Berry Brazelton, "Behavioral competence of the newborn infant," in *Neonatology*, Avery, ed., p. 322.

[7]*Babb v. Murray*, 26. Cal. App. 2d 153, 79 P.2d 159 (1938). For an analysis of this case and *Capelouto*, see Peter N. Kalionzes, "Infant Pain and Suffering: The Valuation Dilemma." *Pepperdine Law Review*, 1, 1973, 102–115.

[8]The case is known as *Capelouto v Kaiser Foundation Hospitals*. This quote may be found at 7 Cal. 3d at 894, 103 Cal. Rptr. at 860.

[9]Eric J. Cassell, "The Nature of Suffering and the Goals of Medicine," *New England Journal of Medicine*, 306 (11), 639–645, at 640.

[10]David H. Smith, "Suffering, Medicine and Christian Theology," unpublished manuscript delivered to the Hastings Center Research Group on Death, Suffering and Well-Being.

[11]*Oxford English Dictionary*

[12]Smith, op. cit., p. 3141.

[13]Joy Hinson Penticuff. Personal communication, August 1983.

[14]Miriam Sherman, "Psychiatry in the Neonatal Intensive Care Unit," *Clinics in Perinatology*, 7 (1), March 1980, 33–46.

[15]Jo-Eileen Gyulay, *The Dying Child*, New York: McGraw-Hill, 1978, p. 35.

[16]Ibid., p. 73.

[17]Leroy E. Loemker, "Theodicy," in *History of Ideas*, Philip P. Weiner, ed. New York: Scribner, 1973, Vol. IV.

[18]Albert Camus. *The Plague*, New York: Modern Library, 1948.

Ethical Principles for the Care of Imperiled Newborns

Toward an Ethic of Ambiguity

John D. Arras

Introduction

The ethical debate surrounding the care of anomalous newborns in the 1980s has been warped by the same strident tone and moral absolutism that led to the impasse over abortion in the 1970s. "Pro-lifers" accuse stunned and agonized parents of "murder most foul" for allowing the deaths of their own children, even as the defenders of "free choice" hurl charges of religious bigotry against those who would merely insist on equal rights for every human being. Each side has definite, unambiguous answers to questions that baffle the average person and elude public consensus: When does "human" life begin? What is the threshold for "meaningful" human life? How weigh the benefits of a child's continued life against the predicted burdens to mother, father, and siblings? To what extent should the protection of nascent human life be a matter of public regulation or private choice?

Although the resurrection of these problems in the present decade attests to how little we have learned from past battles, one would hope that we could at least learn to acknowledge the complexity and ambiguity of the issues, and perhaps even learn to listen to each other a bit more attentive-

ly. No such luck. Although the general public's confusion
perfectly reflects the ambiguity at the core of the "Baby Doe"
controversy,[1] and although some commentators have managed somehow to acknowledge the moral pull of the opposing
briefs, the leading advocates of parental choice and of the
child's rights appear to be equally motivated by an illusory
quest for moral certainty in an area where such certainty
simply cannot be had. In the following discussion I shall try to
articulate a middle way between the polar absolutes of unquestioned parental discretion, on the one hand, and a rigid
invocation of children's rights, on the other.

Substance and Process

As in most areas of ethical concern, we are confronted
here by two distinct issues: a substantive question ("When, if
ever, is it morally permissible to allow impaired newborns to
die?") and a procedural question ("Who decides?"). A crucial
preliminary problem is to determine how these questions
relate to each other. There are some who argue as though the
procedural question were fundamental.[2] They claim that,
since this is an area of intense public disagreement, parents
should be allowed unfettered discretion in making decisions
bearing on the life or death of their children. These "pro-
choice" advocates thus encourage physicians, nurses, and
social workers to support (or at least acquiesce in) even those
parental choices that appear to contravene the child's best
interest.[3]

The unstated assumption animating this pro-choice position is the idea that, at least in this contested area, we need
only decide on a fair and reasonable *procedure* in order to
guarantee *results* that are just and right. This is a good example
of what Rawls calls "pure procedural justice": in the absence
of any independent criterion for what would count as a just
result, we can establish procedural rules that can be agreed
upon as being fair; and we can accept as fair *whatever* results
happen to be generated by our procedure.[4] Examples of this
sort of situation must be rare in the domain of morality; and it

is interesting to note that Rawls chooses to illustrate the workings of pure procedural justice with an example drawn from
the nonmoral sphere: gambling. At the world series of poker,
one player emerges with all the chips; and, so long as everyone has agreed in advance to play by a certain set of rules, this
lopsided result is acknowledged by all to be fair.

Can the advocates of parental discretion seriously maintain that the issue of impaired newborns is really just a problem of pure procedural justice? Of course not. Suppose, for
example, that Mr. and Ms. Smith-Jones are the parents of a
bouncing baby girl, normal in all respects except for a duodenal atresia requiring ordinary surgical repair. The Smith-
Joneses refuse to sign the operative consent on the ground
that they "really were hoping for a boy." Or, to make matters
even clearer, suppose they refuse to sign, explaining, "You
see, Missy here is a Scorpio, but we were really hoping for a
Sagittarius. We know you'll understand and support our decision." The fact that no self-respecting health professional
would accede to such a request indicates that, at least in cases
such as this, we acknowledge independent substantive standards of right and wrong. We know that it would be wrong to
allow this child to die for want of corrective surgery, no matter
what the parents have to say about it. Here the question "Who
decides?" is properly subordinated to the question, "What's
right?"

True, the example of the Smith-Joneses was outlandish
and quite atypical of the sorts of dilemmas that concern us
here. Nevertheless, I contend that even in those cases that do
concern us—e.g., the extremely low birthweight baby or the
Down's child with an esophageal atresia—independent standards of right action should govern our decision-making, *no
matter who is granted the discretion to decide.* As Ronald Dworkin
points out in the context of judicial decision-making, the
notion of "discretion" comes in different varieties and
strengths.[5] When someone is charged with making a decision
in the absence of a governing standard—for example, choosing any two students to carry a message to the principal—the
chooser exercises discretion in the strong sense. Although
one can be criticized for choosing unwisely—say, the two
students with the worst sense of direction—one has not been

given any standard or norm against which he or she must judge the students. By contrast, when someone is told to make a choice in light of a critical standard—for example, choosing the most productive employee—the chooser exercises discretion in a weak sense. The chooser must decide, and the decision may require a great deal of thought and good judgment; yet he or she may still be held accountable for failing to fulfill the mandate. In spite of a most conscientious effort, a judge might overlook the most productive employee. We would say that this decisionmaker was bound by a definite standard, even in cases where it would not be clear to the most careful observer who the most productive employee was. The choice might be extremely difficult, yielding no clear-cut right answers, yet it is still clearly governed by a norm.

Now, I suggest that the kinds of hard choices that perplex us, involving extremely premature and anomalous newborns, are "discretionary" only in this second, weaker sense. Those who decide must exercise discretion—that is, they must carefully deliberate under conditions of great uncertainty—but this does not mean that they are free to decide however they wish, according to whatever standard appeals to them. The mere fact that a certain class of decisions is difficult does not entail the absence of any governing standard (that is, it does not entail "discretion" in the strong sense). I would argue, then, that unless one is willing to claim that the newborn child enjoys no independent moral standing whatever,[6] our first priority should be to identify and explicate those substantive moral standards that ought to govern decision-making for gravely afflicted newborns.

This is not to suggest, of course, that procedural questions are unimportant. It may well turn out that, when confronted by the hard case, our substantive principles will founder on ambiguity and uncertainty. At such a juncture, we may not know what the "right answer" is, and we shall then have to "solve" the problem by giving someone leave to cut the Gordian knot. Even then, however, agreement on substantive principles will be vital, if only to draw the boundaries of *legitimate* discretion.

The Distinction Between "Ordinary" and "Extraordinary" Treatments

One potentially useful guide to decision-making in this area is provided by the well-known, but widely misunderstood, distinction between so-called "ordinary" and "extraordinary" means of treatment. However these terms are defined, those who avail themselves of this distinction mean to say that the use of "ordinary" treatments is *morally incumbent* on physicians, whereas all "extraordinary" treatments are deemed to be *morally optional*. Physicians *may* offer an "extraordinary" treatment to a patient, but they are not morally obligated to employ such measures. The problem, then, is to clarify which treatments qualify as "ordinary" (ethically mandatory) and which as "extraordinary" (ethically optional).

Although these terms enjoy a very wide circulation—one encounters them in court decisions, living wills, on medical rounds, and in the daily press—they are interpreted in radically different ways, and therefore have been the source of a great deal of confusion. For our purposes, all of these varied interpretations can be conveniently lumped together under two very general headings. One approach is to conceive of the "ordinary/extraordinary" dyad as a so-called "categorical distinction"[7] referring to entire classes or categories of treatments. It is commonly said, for example, that respirators or complicated surgery are "extraordinary means," whereas intravenous feeding or antibiotics are "ordinary." It is as though the words were intended to function as labels attaching to specific kinds of treatments, quite apart from their respective effects on particular patients.

One problem with this approach is that different people end up attaching different labels to the same treatments. In the famous Quinlan case, for example, Mr. Quinlan argued eloquently that the respirator (seemingly) necessary to sustain his daughter's life was an extraordinary means; whereas Karen's physicians and legal guardian felt that a respirator was a part of ordinary medical practice. It appears likely that

Mr. Quinlan would define as "extraordinary" all those treatments that involve highly artificial or intricate technologies; at
least this is a reasonable inference from the report that he
balked at the suggestion that Karen's intravenous drip also be
discontinued.[8] Since the physicians could not quarrel with the
judgment that a respirator is the quintessential "high tech"
medical device, they must have had other criteria in mind. For
them, "ordinary" must have meant something like "usual" or
"commonplace" in the contemporary practice of medicine.
Since the physicians felt morally obligated to utilize all "ordinary" measures, they felt morally compelled to keep Karen on
the respirator, just as Mr. Quinlan felt compelled to continue
the intravenous drip.

The weakness of Mr. Quinlan's approach becomes apparent as soon as we realize that a treatment's degree of
artificiality or technological complexity has little to do with the
moral issue posed by its potential use. A patient undergoing
open heart surgery will require the use of an artificial respirator in the recovery room; she requires artificial breathing
assistance precisely because her natural capacity is temporarily impaired. Were a benighted physician to withhold the
respirator and thereby cause the death of such a patient, it
would be a lame excuse indeed for him to say, "Well, the
respirator *is* a highly complex and artificial technological device: so doesn't that make its use morally optional?" Clearly, it
is not a treatment's status as "high-tech" that renders it morally optional; rather, it must have something to do with how the
treatment will impinge on the health and well-being of the
patient.

The physicians' criterion of ordinary and extraordinary
means is subject to the same criticism. Apart from the obvious
fact that their criterion of ordinariness is disturbingly relativis-
tic—today's extraordinary or heroic venture routinely becomes tomorrow's scut work—it is hard to see what ordinariness in this merely statistical sense has to do with the *moral*
issue. Why, in other words, should the mere fact that physicians habitually use a certain treatment render it morally
obligatory in particular cases? Were Karen's doctors under a
moral obligation to keep her on the respirator merely because
such devices have become part of the warp and woof of

modern medical practice? Again, posing the moral issue points us beyond this categorical or labeling approach to the distinction towards an inquiry into the consequences of treatments for particular patients.[9]

The rich tradition of Catholic moral theology provides an alternative rendering of the distinction between ordinary and extraordinary means, one that avoids the misleading abstractness of the categorical approach. I hope that I do not do too much violence to the Catholic understanding of the distinction by uprooting it from its traditional theological moorings and inserting it into this secular context. In any case, my aim here is to find useful clues to our own inquiry, not to provide an accurate account of the Catholic position.

The crucial feature of this "Catholic" gloss on the distinction is its commonsensical emphasis on the benefits and burdens of any proposed therapy for particular patients: "*Ordinary* means of preserving life are all medicines, treatments, and operations, which offer a reasonable hope of benefit for the patient and which can be obtained and used without excessive expense, pain, or other inconvenience. . . . *Extraordinary* means of preserving life . . . mean all medicines, treatments, and operations, which cannot be obtained without excessive expense, pain or other inconvenience, or which, if used, would not offer a reasonable hope of benefit."[10]

Thus, a treatment may be deemed "extraordinary" if it meets either of the following conditions. First, a treatment may offer no reasonable hope of benefit to a particular patient in specified circumstances; that is, it might be *useless* in terms of producing a desired effect. For example, vasopressors may prove useless on a patient dying from inoperable heart disease. At best, they will raise the patient's blood pressure for only a short time, merely prolonging the patient's dying. They will not cure the patient or reverse the terminal condition. Or there may be situations in which a proposed treatment holds out *some* hope of benefit, but the odds of a successful outcome are so remote that opting for such a treatment will be deemed unreasonable. (This will be especially true when, in addition to offering a remote prospect of success, a treatment will impose predictable burdens on a patient.)

This should not be interpreted, however, as a call to abandon the dying patient. The uselessness of curative measures does not imply the uselessness of all sorts of other measures, whose primary purpose might be to comfort the dying patient.[11] Thus, while the aggressive use of antibiotics, dialysis, vasopressors, and surgery may well be useless, and therefore extraordinary, in the case of a dying heart patient, those treatments geared to comforting the dying patient—such as the intravenous drip, suctioning, bathing, hand-holding, and so on—may well be morally imperative. Even these latter treatments, however, can *become* morally optional if the dying patient's condition (e.g., permanent coma) places her beyond the pale of human comforting.

Second, a treatment may be deemed "extraordinary" if the benefits that it offers a patient are outweighed by the burdens that it will impose—in short, if it is *excessively burdensome*. Treatments may be foregone because they are too painful, too risky, or too inconvenient, even if they are not completely useless. For example, an elderly patient may benefit from a few additional weeks, months, or even years by submitting to regular kidney dialysis; but if the patient tolerates dialysis poorly, experiencing severe pain, discomfort, and nausea, he or she may reasonably decide that the burdens of continued existence are too great. Those working within this "Catholic" tradition would say that in cases such as this one, where there is a lack of proportion[12] between benefit and burden, the patient is not morally obligated to accept treatment, just as physicians are not obligated to urge its acceptance.

Extraordinary Means and Quality of Life

Before we apply this interpretation of the ordinary/extraordinary distinction to the case of the imperiled newborn, one additional comment is in order. This observation concerns the relevance of "quality of life" judgments for determining which medical treatments are extraordinary. At first glance, it would seem that certain treatments could be deemed extraordinary in certain contexts without having recourse to judgments bearing on the *quality* of a patient's life.

Given the genesis of this distinction in Catholic moral theology, this is precisely what one would expect. At the risk of oversimplifying somewhat, one could say that the distinction between ordinary and extraordinary means was devised to reconcile a firm commitment to the sacredness of each individual life with the belief that our biological existence need not be prolonged indefinitely at great cost (physical, psychological, or perhaps even financial) to the patient. The moral category of extraordinary means allows us to forego certain treatments without submerging our commitment to the individual in murky comparative judgments concerning the "quality" of life.

Now, it is true that *some* advocates of "quality of life" reasoning contend that patients with a very low quality of life thereby become "less human" or less *worthy* of care on account of their loss of capacity, and it is precisely this kind of derogatory judgment that the Catholic tradition has attempted to exclude. However, the phrase "quality of life" need not carry this unfortunate connotation. It could, for example, simply denote the undesirable nature of certain states of body or mind, regardless of our commitment to the sacredness of each individual. That is, we could affirm the view that each individual is of inestimable worth as a human being, but still make judgments bearing on the experiential quality of a person's life. Thus, we can say that someone is happy and flourishing, or that he is depressed and racked with pain. The point I wish to make here is that if we understand "quality of life" in this weaker sense, then many determinations of extraordinary means will *necessarily* be based upon quality of life judgments.

The inevitability of this conclusion becomes evident when we focus our attention on the notion of "excessive burden."[13] The question is this: how are we to know when a treatment becomes *excessively* burdensome—i.e., excessively risky, painful, costly—in the absence of any reference to the quality of future life that the treatment promises to purchase? We can, of course, know that a treatment is painful *tout court* without recourse to our future state of being—it either hurts or it doesn't—but we cannot say that it is excessively painful unless we are able to judge that the quality of life afforded by the treatment is or is not "worth the trouble." For example, a 95 year-old man may decide to forego painful surgery to restore the full range of movement to his shoulder, preferring life with limited movement to the pain of surgery; but a Jimmy Connors at the peak of his tennis career would no doubt consider the same operation to be absolutely essential to his future life. There is thus no way to judge a treatment to be excessively burdensome without considering the quality of life that the patient can look forward to. If the burdens usually imposed by a treatment promise to outweigh the benefits, the treatment is "extraordinary."

Applying the Distinction to Dying Newborns

Coming finally to the issue at hand, how are we to apply the notion of extraordinary means to the impaired newborn? The case of the *dying* infant, though undoubtedly the most poignant, is theoretically unproblematical. If an infant is reliably diagnosed as having an incurable and fatal condition, and if he or she is reliably judged to be in the very process of dying, then any and all curative measures ought to be deemed *useless* and set aside. Surgery is thus an extraordinary means for a child born with an abnormally small left ventricle—a fatal and inoperable condition.

While all the other fronts of the anomalous newborn controversy are strenuously contested by opposing forces, consensus *has* been reached here: in the case of the dying newborn, everyone agrees that measures originally designed to cure may merely prolong the process of dying, and therefore may be ethically foregone. Even "pro-life" theologians, such as Paul Ramsey,[14] and "pro-life" physicians, such as Dr. C. Everett Koop (who is also President Reagan's Surgeon General),[15] see no good reason to press useless curative measures on dying patients. For them, there is a significant difference between infant *euthanasia*—i.e., intentionally bringing about the death of an infant by either active or passive means—and withdrawing useless treatments in the face of death's inexorable advance. In foregoing useless treatments we do not intend to kill—the child's underlying affliction is doing that—rather, we withdraw nonbeneficial cures so that the child may be spared a needless prolongation of his death agony. Here and here alone, say the "pro-life" advocates, when the patient has irrevocably entered into the process of dying, does the bare distinction between killing and "allowing to die" mark a significant moral difference.

Although curative means may thus be foregone in the case of dying infants, measures designed merely to comfort such children in their dying *can* often benefit them. When this is the case—for example, when food, intravenous drips, suctioning, or caressing can make a difference—such measures should be taken. Even though we must often give up trying to save certain infants, they should never be abandoned. As Ramsey emphasized, when cure is no longer possible, *caring* for the dying patient assumes paramount importance.[16]

In spite of the commonsensical appeal of this approach, a thorny problem of categorization ought to preclude a mechanical application of the notion of extraordinary means to the dying newborn. Who counts as a *dying* newborn? It might make perfect sense in the abstract to say that the dying child need not be kept alive with useless curative measures, but in the concrete situation of the newborn nursery, it is often far from clear whether a particular patient ought to be placed in this category. How are we to know when a child has entered into the process of his or her own dying? What are the distinguishing marks of the dying patient?

Although this problem will strike many "pro-choice" readers as a mere semantic quibble, its resolution should be a matter of pressing urgency for "pro-life" theorists. Those who would sanction nontreatment on quality of life grounds have the "luxury" of dismissing this as a "mere" problem of labeling because, when confronted with a borderline case, they have no qualms about switching from a "useless treatment" analysis to the "excessive burden" prong of the extraordinary means test. For them, the problem of correctly categorizing the dying patient lacks urgency because, whether or not a child is actually in the process of dying, they can always ask, "Will our continued ministrations impose excessive burdens?" Strict "pro-lifers" do not have this "luxury." For them, the problem of correctly applying the concept "dying patient" should assume paramount importance because, as we shall soon see, they emphatically reject the imposition of the "excessive burden" test on patients who lack the capacity to decide for themselves whether any particular treatment is "excessive." Limited as they are, then, to the "useless treatment for dying patients" test, "pro-life" advocates ought to view the proper identification of the dying patient as an urgent task fraught with practical consequences.

Unfortunately, the "pro-life" theorists have not yet risen to the challenge: they have failed to provide us with a set of conditions that all "dying" newborns must satisfy. What they *have* done, however, is to exclude certain likely candidates from this category, thereby restricting considerably the reach of a policy that would "allow the deaths" of anomalous newborns. In particular, they have explicitly rejected the idea that treatments may be withheld from newborns who are *merely* adjudged to be terminally, incurably ill.[17] Thus, an infant diagnosed as having Tay-Sachs disease—an incurable and fatal condition affecting the nervous system—would not count as a dying patient in the early stages of the disease, even though death inevitably comes within three or four years.

Apart from this refusal to identify "dying" with the merely incurable or terminal, "pro-life" advocates have done little to clarify the boundaries of the concept. Apparently convinced that the elaboration and application of the concept "dying patient" is strictly a matter of *medical* expertise, Paul Ramsey has not even attempted any further clarification of the concept. Distinguishing between the merely incurable and the truly dying, Ramsey states, is simply a matter of medical judgment.[18]

Now, whether this failure to go farther stems from an acknowledged lack of appropriate expertise, or rather from an unacknowledged reluctance to tread on logical quicksand, the resulting lack of clarity is disconcerting. We may have no trouble concurring in the doctors' judgment that the infant born with an absent left ventricle is in the process of dying; but what of the Trisomy-13 baby—born blind, microcephalic, and destined to die within a matter of months? Or what of the nearly anencephalic baby, born without a higher brain, subsisting on shards of lower brainstem activity? In each case, the child is doomed to expire within a matter of weeks or months.

But whether the child dies in the newborn nursery or eight months later will often depend on how aggressively the child is treated. Should we say that such children have entered into the process of dying, or that they are (merely) incurable but not yet dying? Do they bear a greater resemblance to the baby with the absent ventricle, or to the baby with Tay-Sachs? If a child's *time remaining* is the crucial factor distinguishing the dying from the (merely) incurable child, why should three years mark a significant difference, but not eight months? Finally, whether or not doctors deem eight extra months of life sufficient to distinguish the Trisomy 13 baby from the dying newborn, is this assessment a matter of purely medical expertise, or does it rest on a nonmedical value judgment?

Applying the Distinction to the Nondying Infant

When we pass from the case of the dying infant, however we choose to define that condition, to the case of the nondying anomalous child, public consensus falls apart and the present debate begins in earnest. What to do with the extremely low birthweight child with an intracranial bleed or with the child born with spina bifida, Down's syndrome, Tay-Sachs, or hemophilia? Often these infants will present with medical conditions requiring some sort of life-sustaining treatment: the plight of Baby Doe, a Down's syndrome child born with an esophageal atresia, is a classic example. The proposed treatment, surgical repair of the atresia in Baby Doe's case, will permit the child's continued existence, but it will have no effect whatever on the underlying congenital defect. Were Baby Doe to survive, it would have been as a child, and subsequently as an adult, afflicted with Down's syndrome. Is it ever morally permissible to withhold medical treatment from such infants who are not in the process of dying?

A "Medical Indications" Policy

According to many concerned moralists and the Reagan administration, there are no compelling reasons to withhold life-sustaining care from nondying newborns. Acting on the

legislative authority of Section 504 of the Rehabilitation Act of
1973, which forbids discrimination on the basis of handicap by
any federally funded programs,[19] the present administration
has recently attempted to enforce a policy mandating the
provision of "medically indicated" treatments to all anomalous newborns.

Significantly, this policy does not require the impossible:
it specifically excludes from its scope the case of the infant
born terminally ill, for whom medical care will only prolong
the process of dying.[20] But for all other infants, the administration is intent on imposing what Paul Ramsey has called a
"medical indications policy."[21] This means that any and all
treatments (not to mention nutrition) deemed "medically indicated" must be provided to all impaired newborns. If a
proposed treatment will "help" an infant or prolong his life, it
must be administered. In most cases, the existence of a "handicap"—i.e., mental retardation, malformed features, and so
on—is not a legitimate *medical* consideration; however, when
a physician concludes that the handicap itself renders treatment medically contraindicated—for example, an especially
large spina bifida lesion judged not amenable to surgery at a
particular time—she is not required to treat.

The positive intent of the administration is thus to require
all medically indicated treatments without intruding on the
exercise of *legitimate* medical judgment. The express negative
intention of the policy is to prevent doctors, hospital adminis-
trators, and parents from invoking "subjective" quality of life
considerations regarding the care of anomalous newborns. So
long as an infant is not dying, he must be treated, no matter
how poor his predicted future quality of life. An obvious
corollary of such a policy is the exclusion of the "excessive
burden" prong of our definition of extraordinary means.
Since, as we have seen, treatments cannot be judged to be
excessively burdensome without some reference to the quality
of expected future life, the Reagan policy must necessarily
restrict the meaning of "extraordinary means" to those treatments that prolong the dying process. So long as a treatment
is medically indicated—no matter what burdens it threatens
to impose on children, their families, or society—it must be
given to an impaired child.

Although the "Section 504 strategy" meshes nicely with the administration's well-known sympathies for the conservative "Right-to-Life Movement," Reagan's advisers chose to mount this conservative policy on a peg well-worn by liberal causes: viz., the nondiscrimination principle.[22] This principle says that, in the allocation of benefits, it is unjust to "discriminate" against people. Justice demands, not that everyone be favorably affected by a distribution, but rather that no one be denied benefits on the ground of characteristics that are *irrelevant* to the purposes of the distribution in question. In order to make out a case of unjust discrimination, one must show not only that a particular distribution favors some people over others, but also that it does so for morally irrelevant reasons. Thus, just as a child's religion or race are irrelevant to the question of whether it should receive needed health care, so too are its blindness, mental retardation, or meningomyelocele. Just as black and white persons are invested with equal moral worth in the eyes of our public ethics and law, so the anomalous child is said to be the moral equal of the normal child. Now, since justice demands, through the nondiscrimination principle, that like cases be treated alike, and since anomalous and normal children are "alike" in being moral equals, justice demands similar treatment for them both. If an otherwise normal child would be treated for a certain condition—say, an esophageal fistula—then we cannot deny the same treatment to a Down's syndrome child, merely because it is a Down's child, without thereby violating the nondiscrimination principle.

There is no denying that the Reagan administration's policy will dictate the morally correct result in many cases. There has been widespread discrimination against mentally and physically handicapped persons in our society, and any practice that has the effect of denying life-sustaining nutrition or medical care to this morally "suspect classification"[23] must be viewed with extreme prejudice. This is especially true in the case of anomalous newborns, where many factors may conspire to yield unjust treatment. The parents of such children are often called upon to make life or death decisions in the midst of severe psychological trauma and in the face of great uncertainty regarding both the nature of their child's affliction and its long-term prognosis. It is not unreasonable to think that many otherwise good and caring parents might allow their shock, disappointment, and fear to dictate decisions prejudicial to the interests of their anomalous children. For their part, the children themselves are voiceless. Unlike their handicapped brothers and sisters who lobbied and fought for the Federal Rehabilitation Act of 1973, and who wage a daily struggle against societal neglect and ostracism, the handicapped newborn cannot plead its own case for fair treatment. Given the likelihood of parental conflict of interest and the absence of a forceful advocate for the child's best interests, it is not surprising, and by no means morally suspect, for the present administration to attempt to intervene on behalf of handicapped newborns.

Indeed, the necessity for *some* response, whether public or private, to the vulnerability of the anomalous child has been amply demonstrated by the *cause célèbre* of Baby Doe.[24] Although the Indiana court's records have been sealed off from public scrutiny, it appears that Baby Doe suffered from no other infirmities than his Down's syndrome and an esophageal atresia. Although the atresia was amenable to routine surgical repair, the same could not be said of the child's underlying mental and physical defects that constitute Down's syndrome. I think it can be fairly concluded that Baby Doe was "allowed to die" (i.e., killed) simply on account of his handicap. This was evidently also the conclusion of the Reagan administration because two weeks after hearing the news of Baby Doe, the President ordered the Department of Health and Human Services to protect future Baby Does by means of a strict "medical indications policy." In order to facilitate compliance with this new federal policy, the President established a toll-free "Baby Doe Hotline" on which to report perceived violations.[25]

No matter what the shortcomings of the administration's *policy*—which, as we shall soon see, are legion—the President's moral response to the Baby Doe incident was, I would argue, perfectly appropriate. The question, however, is whether the policy inspired by the Baby Doe case is good *policy*, not whether it was well-motivated, or even whether it will yield correct results in similar cases. I shall argue that the

Infant Doe rule is very bad policy, and that it is bad precisely because it was tailored to the case of Baby Doe.

Simplistic Social Policy

By attempting to erect an entire social policy on the slender reed of a few miscarriages of justice, the Reagan administration burdens the American people with a hopelessly *simplistic* rule. As we shall see, the Infant Doe Rule oversimplifies an extraordinarily complex moral and factual situation. Adequate social policy needs to be formulated on the basis of a broad understanding of moral and medical realities, not on moral outrage directed, no matter how appropriately, against a few highly publicized cases.

Medical Complexity

In the first place, the Infant Doe Rule is premised upon an overly simple picture of the medical context. By implicitly assuming that all congenital defects are as potentially "benign" as Down's syndrome, and that all contested treatments are as clearly beneficial as the routine repair of an esophageal fistula, the Infant Doe Rule accomplishes a Herculean feat of oversimplification. The truth of the matter is that there is a vast range of conditions that can imperil a newborn's life and health. Some are relatively benign, like mild Down's syndrome and low-lesion spina bifida, whereas others are catastrophic in the sense that they portend a very short life span with virtually no cognitive or affective capacities. Matched against this spectrum of illnesses is a highly variegated spectrum of possible treatments, ranging from simple oral feeding to invasive surgery and chronic kidney dialysis.

In order fully to appreciate the complexity of the medical context enveloping the Baby Does of this country, consider the options confronting the caregivers of a Trisomy-13 child. This is a child with microcephaly, an abnormally small brain, portending a near-total absence of the abilities to think and communicate with other human beings. The child will almost certainly die within a year, even if maximal therapeutic efforts are made on the child's behalf. What does it mean to give equal concern and respect, or to discriminate against, such a child? If the child cannot secure sufficient milk from a plastic nipple, should the nipple's opening be enlarged? If it develops an infection, should it be treated with antibiotics? In case of breathing difficulties, should we assist breathing or place the child (perhaps permanently) on a respirator? Should heart defects be diagnosed through invasive catheterization and repaired by major surgery? If the kidneys fail, do we resort to chronic dialysis? Finally, supposing that the child's heart suddenly arrests, should a full-scale attempt be made to resuscitate it?[26]

As this bewildering list of questions indicates, the problem of caring for imperiled newborns is not usually an *all or nothing* dilemma between so-called "customary care" on the one hand, and doing nothing on the other. Rather, and this is especially true for the very low birthweight infants, it is a question of how far we should go, given this infant's condition, prognosis, and the relative benefits and burdens of the proposed treatment. Against the backdrop of this kind of medical complexity, the administration's nondiscrimination principle is more likely to function as a bioethical *nostrum* than as a useful guide to conduct in the neonatal nursery. H. L. Mencken said it best: "For every human problem there is a solution that is simple, neat, and *wrong*."

The Concept of Discrimination

A second major feat of oversimplification wrought by the Infant Doe Rule lies in its operative notion of "discrimination." Although the Department of Health and Human Services presents the rule as "an equal treatment, nondiscrimination standard,"[27] and although every well-meaning citizen would no doubt condemn the practice of *unjust* discrimination against the handicapped, the great difficulty facing us is to determine what constitutes discrimination against gravely ill newborns. To state the problem in a more positive vein, how do we grant equal concern and respect to persons born with severe congenital anomalies or other catastrophic impairments?

We already know in a rough and ready way what counts as discrimination in the areas of race and sex discrimination; and we also have an intuitive sense of what our lawmakers must have had in mind when they passed Section 504, outlawing discrimination against handicapped persons in areas such as employment, education, and access to public facilities. But what is the meaning of discrimination in the neonatal nursery?[28] The various incarnations of the Infant Doe Rule suggest three possible interpretations.

Failure to Provide Customary Care

The first and most problematic gloss of "discriminatory nontreatment" conceives of it as a failure to render so-called "customary care." The poster that DHHS had originally designed for placement in all neonatal units stated in relevant part: "Any person having knowledge that a handicapped infant is being discriminatorily denied food or *customary* medical care should immediately contact. . . ."[29]

Although the notion of providing "customary care" supplies a workable legal standard of negligence in most areas of life; and although it might well provide us with sufficient guidance in the easy neonatal cases—for example, mandating the surgical correction of an esophageal atresia in an otherwise normal baby; the standard of customary care is singularly unhelpful in the hard cases that concern us here. This is because, quite simply, there exists no such standard of care governing the hard cases. Some physicians think that all imperiled newborns should receive maximal treatment, whereas others are willing to leave it up to the parents when the benefits of intervention are unclear. Indeed, if there were a standard of customary care in such cases, we would not be debating the issue so strenuously on medical rounds, in the press, and in the courts.

Equal Treatment

A second interpretation of discrimination advanced in the explication of the rule is based on the idea of equal treatment. According to this view, a person or institution is guilty of discrimination if a handicapped child is deprived of care that would have been provided to an otherwise normal child in similar circumstances.[30] Just as one may not discriminate against persons on the basis of race, color, or creed, so one may not discriminate against a person *solely* on account of his or her handicap.

Significantly, this standard works very well in factual situations closely analogous to the original Bloomington Baby Doe case. That is, when there is no difference between two patients—except for the fact that one will always be retarded, because of Down's syndrome, and the other is mentally normal—it strikes most people today as unjust to allow the former to die *solely* on the ground of her handicap. Mental retardation, according to this emerging social consensus, is not in itself a morally significant difference between the Down's baby and the otherwise normal child.

In spite of its initial plausibility, this "unequal treatment" interpretation of discrimination also founders on the difficult cases. Specifically, an equal treatment gloss is inapplicable when a given defect, by definition, is not present in "otherwise normal" infants. Although an "otherwise normal" child might well have an esophageal or duodenal atresia, it does not have meningomyelocoele, anencephaly, or Trisomy-13. In such cases, it is meaningless to ask, "Apart from this child's hydrocephalus and possible retardation, would we seal up the spinal column of an otherwise normal baby?"

This equal treatment standard founders more dramatically and more interestingly when a child is burdened with an anomaly or deficiency of such magnitude that comparisons with normal children seem utterly beside the point. We may all agree, for example, that a Down's child born with a cleft palate should have corrective surgery in spite of its retardation. But what if, instead of Down's syndrome, the child was born with holoprosencephaly (a devastating malformation of the brain involving the absence of the entire forebrain), microcephaly (an abnormally small brain), and cebocephaly (the absence of a nasal passage)? Given such catastrophic deficits, pediatricians can confidently state that this child lacks the intellectual capacity to communicate with other persons in a meaningful way and that, in any case, the baby is doomed to die a premature death. They note, however, that very aggressive surgical, medical, and nursing interventions could *possibly* sustain the child for a year or longer. (That is, the child cannot be routinely classified as a "dying" patient.)

Surely, if an otherwise normal child were to be born without a nasal passage, no responsible physician or loving parent would suggest for a moment that the air passage not be surgically created. (This is a life-sustaining procedure because newborns cannot breathe through their mouths.) But this reasoning does not carry over in anything like an automatic or intuitively obvious way to the hypothetical case at hand. Here it is a difficult and agonizing question whether the child's temporary breathing tube should be replaced by a surgically created nasal passage, but whether we operate or not, the child's holoprosencephaly, microcephaly, and consequent lack of capacity for human interaction all appear to be relevant factors in our decision-making in a way that the other child's Down's syndrome was not. (An even more dramatic example would be provided by the case of a *nearly* anencephalic baby, subsisting on remnants of brainstem activity, who required a similar operation. The infant is not "dying": aggressive care could keep it going for a long time. Would the operation offer the baby a "medical benefit"?) My point, then, is simply this: If the degree of severity of a child's "handicap" should matter in such cases, then no standard mandating the "evenhanded treatment" of handicapped and otherwise normal infants will prove adequate to the complexity of these extreme and unhappy situations.

No Quality of Life Judgments

The third interpretation of "discrimination" goes to the theoretical core of the Infant Doe Rule. According to this reading, it is discriminatory to deny medical benefits to a child "solely on the basis of an infant's present or anticipated mental or physical impairments."[31] In other words, it is discrimination to withhold nutrition or medical care on "quality of life" grounds. So long as a child is not embarked on the process of his own dying, the Infant Doe Rule is explicit: All such children must be treated, no matter how poor their present or future quality of life. So long as the infant will "benefit" from a given treatment—a supposedly "medical" judgment—the Down's baby must receive its thoracic surgery, the spina bifida baby must have its spinal lesion surgically closed, and every child must be fed and cared for.

I contend that this philosophical rejection of quality-of-life reasoning, rather than any well-thought-out moral theory of nondiscrimination, constituted the theoretical nerve of the Infant Doe Rule. The language of nondiscrimination and the trappings of civil rights law appear merely to have served the administration as convenient vehicles for advancing a "sanctity of life" philosophy in this debate. Notwithstanding this possibility that the administration's interest in the concept of discrimination was purely instrumental, we can still profit from a closer look at this problematic concept. Indeed, I shall argue that the administration's nondiscrimination arguments fail for precisely the same reason that any rigid "sanctity of life" position will fail—that is, both kinds of argument fail to demonstrate why quality of life considerations ought to be discounted by the caregivers and parents of severely impaired newborns. In the end, both of these strategies merely assume their identical conclusion.

What, then, is discrimination? In its most general and nonmoral sense, to discriminate is simply to make distinctions between two or more different things or persons. We say, for example, that a wine connoisseur possesses "discriminating" taste—that is, that he or she knows how to distinguish good wine from bad. Now, if discrimination in this most general sense refers simply to making distinctions, what is it that renders certain discriminations ethically unjust? We can begin by noting that it is *not* the mere fact of unequal treatment. Although morally invidious acts of discrimination will often involve unequal treatment, not all unequal distributions are discriminatory in this sense. The applicant who is rejected by a law school because he or she cannot read or write has not been the victim of discrimination. Why? Because such skills as reading and writing are highly *relevant* to the enterprise of selecting students for legal study. However, if a female applicant were refused admission solely on the ground of her sex, we would say that she has indeed been discriminated against. Surely, we would say, her sex is irrelevant to the business of studying or practicing law. Hence, sex should not be used as a basis for denying anyone a seat in a law school.

The key to the notion of discrimination in this morally pejorative sense, then, is the relevance or irrelevance of a given characteristic or criterion to the matter at hand. When a distribution is based on a relevant trait—e.g., considering only females for a position as wet nurse, or only individuals with excellent sight for the job of jet pilot—there is no discrimination; but when people are treated differently solely on the basis of irrelevant criteria, then we have genuine discrimination.[32]

Returning to the issue of gravely impaired newborns, we are now better equipped to define discrimination in the context of the neonatal nursery. We can now clearly see that the fundamental question is *not* whether an "otherwise normal" child would have received a certain treatment; rather, the question is: Are there any morally significant differences between certain imperiled newborns and their normal counterparts that might justify nontreatment?

The Best Interests of the Child

One important and plausible candidate for a morally relevant distinction between normal and certain anomalous newborns resides in the notion of their respective "best interests."[33] Continued life is obviously in the interests of the normal child, but what of the child impaired by *profound* brain damage, whose days will be measured by operations, and whose pain will be unrelieved by the communication of human sympathy? Would it not be in this child's best interests to die, rather than endure a life of meaningless suffering? And if so, would that not provide us with a morally relevant distinction?

The case for allowing the deaths of anomalous newborns on the ground of their "best interests" is by far the most compelling argument for infant euthanasia. By focusing on the *child's* interests, this standard avoids morally dubious utilitarian justifications based on the well-being of other interested parties, such as parents, siblings, or even of society at large; and by giving *full weight* to the child's interests, this approach can justify nontreatment in certain cases without having to make the (usually pernicious) assumption that the child is a nonperson with no standing in the human community. We shall soon see, however, that, in spite of these advantages, the best interests standard presents us with staggering problems of interpretation and application.

A Preliminary Objection

Before coming to these problems, we must confront a direct challenge to the very legitimacy of the best interests standard itself. Certain theologians and jurists who oppose any resort to "quality of life" judgments on behalf of voiceless patients have consistently argued that the best interests standard forces us to make an impossible comparison. Although we can and should compare better life against worse life, they argue, we cannot compare the advantages of life, even with severe defects, against the state of nonbeing initiated by death. Since all of our everyday evaluations take place against the backdrop of the assumed value of life, it is said, we cannot make the comparison of life against nonlife demanded by the best interests standard without foundering on conceptual incoherence.[34]

Although this is certainly not the occasion to launch a lengthy metaphysical explanation of how a state of nonexistence might be "preferable" to certain states of existence—how can someone be better off dead?—two points can be made in response to those who claim that death can *never* be in the best interests of a nondying child. First, we can concede the point that a belief in the desirability of life grounds our everyday judgments of "better" and "worse," while still maintaining that the moral choices posed by the plight of the defective newborn are anything but ordinary. Although we usually profess an implicit faith in the conceptual connection between our notions of "life" and "good," the presence of the profoundly anomalous and severely brain-damaged child shatters this everyday faith and severs the assumed connection. In such cases, it is not at all obvious that mere life remains a good to the afflicted person; rather, it seems as though the child's illness has eclipsed the possibility of enjoying those normal human goods that ordinarily predispose us to say that life is good.[35] Moreover, though the existence of severe defects threatens to sever our everyday connection between "life" and "good," another item of everyday faith remains unshaken: We continue to believe that pain and suffering are evil, and that beneficence largely consists in abolishing, or at least ameliorating, these twin scourges of humankind. If an act of mercy killing were to deliver a child from a life of unmitigated pain—if only to a state of non-existence rather than beatitude—it would take a profoundly dogmatic temperament to deny that death was in the child's best interest. Opponents of infant euthanasia could perhaps find other reasons to oppose such an act—for example, the belief that human beings have no *right* to dispose of their own or, *a fortiori*, anyone else's life—but merely on the grounds of individual well-being, there is really no denying that death would be better than a life of intense and constant pain.

Secondly, if it really were impossible for us to detach ourselves from our everyday grounding in the belief that life is good, if we really could not wish someone else dead for his or her own sake, then the same difficulty would perforce attach to first-person judgments as well. If it were *in principle* impossible to say that death would serve the best interests of any child, then it would also be equally impossible for any rational adult to conclude that, given a bleak prognosis and the prospect of burdensome treatments, he or she would be better off dead. I doubt that the more thoughtful opponents of euthanasia would wish to embrace such a result. Even conservative theologians such as Paul Ramsey restrict their rejection of quality of life reasoning—and hence their rejection of "excessive burdensomeness" as a test of extraordinary means—to the class of voiceless patients who have not developed a value system that would allow them to assess the quality of future life for themselves.[36] Ramsey explicitly grants to the competent adult the right to refuse treatments that would render continued life a burden. In order for him to assume this stance, he must admit the *possibility* that death would be in the best interest of certain persons; but this admission pulls the rug out from under the claim that, in principle, we cannot judge death to be in a child's best interest. The most that can be said is that no one has the right to make such judgments, or that allowing adults to make them for children will result in mistaken choices or downright abuse. We cannot say that, in principle, life (even with crushing impairments and constant pain) cannot even be compared with death. At best, such a denial is an unfortunate overstatement; at worst, it could be placed in the service of an overbearing medical paternalism. Indeed, what better way to thwart the choices of mature adults than to deny them the very capacity rationally to choose between an earlier death without treatment and a life of suffering with treatment?

In sum, we can say that death is not always and everywhere the greatest evil that could befall a person. Afflictions of the body and mind might conspire to render continued life a burden for certain individuals, for whom death might be viewed as a benefit—or at least as the lesser of two evils. As the President's Commission recently concluded, the best interests of the child—rather than a spurious nondiscrimination principle—should be our guide. Normally, of course, treatment will clearly be in a child's best interest. But the Commission granted the possibility that some permanent handicaps might be "so severe that continued existence would not be a net benefit to the infant."[37] Our next problem, then, is to determine with more precision how a best interest standard might be applied and to discuss some very weighty problems of interpretation.

Prognostic Uncertainty

A few words are in order about a problem that will plague efforts to implement *any* proposed criterion of nontreatment: *viz.*, the problem of medical and prognostic uncertainty. Since much has already been written elsewhere about the moral implications of medical uncertainty, I shall be very brief here.[38] The problem is this: Decisions must often be made regarding life or death in the absence of reliable data regarding a particular child's future life prospects. No matter where we draw the line separating "meaningful" from "excessively burdensome" life, we will often have to decide under conditions of great uncertainty. The spectrum of neonatal disabilities is very wide, as is the spectrum of disability *within* particular disease categories. The mere presence of Down's syndrome or of spina bifida cystica does not by itself indicate how disabled a child will be. Some Down's children are *profoundly* retarded, though others end up being sufficiently intelligent to live at home with their families or in group homes and eventually to enjoy work in a sheltered shop. The problem is that, except for a few classes of disease, physicians cannot accurately predict the degree of a child's eventual impairment.[39] In the absence of reliable prognostic indicators of future quality of life, the rationality of nontreatment decisions for defective neonates is bound to be diminished. Any medical ethic that places a great value on human life must look with suspicion upon nontreatment decisions taken in the absence of reliable medical information.

Although the persistence of medical uncertainty threatens to undermine the application of *any* set of criteria for nontreatment based on quality of life, the best interests standard is subject to further difficulties of interpretation. It might well turn out that, all things considered, death might serve the best interests of a severely afflicted infant; but before we can reach this conclusion, two important questions must be resolved. First, we must know what burdens to the child are admissible in our moral calculations of benefit and burden. Do any and all burdens to the child count, or only certain kinds of burden? Second, we must have a clearer idea of the point of view from which these burdens will be assessed. Shall it be the viewpoint of the defective child or of the normal adult?

Which Burdens Shall Count?

So far, we have been concerned to catalog the list of physical ailments that might befall a child and render excessively burdensome a prolongation of its life. Now we must consider whether burdens attributable to the child's socioeconomic status can be counted alongside its strictly medical burdens as contributing to an excessively burdensome life and, if so, whether such burdens morally *ought* to be so counted.

In addition to suffering from the retardation associated with Down's syndrome, the repeated surgeries associated with spina bifida, or the iatrogenic toll of respirator therapy for premature infants, a child may suffer from a host of non-medical problems. It may, if allowed to live, suffer from parental rejection. Plagued by frustration, guilt, and disappointment, parents can and do neglect their anomalous children—or, worse yet, abuse them. Or the parents may love and wish to care for their defective child, yet be too ill-equipped to care for it at home and too poor to place it in an acceptable institution. Such parents face the dilemma of keeping their child at home, where the demands of caring for its defects are likely to drain the family of its economic and emotional resources, or handing her over to an institution that is likely to be underfinanced and understaffed. At such an institution, the severely defective child is likely to be relegated to some back ward where it will suffer from lack of love, human contact, and even basic care.[40] "Clearly," it might plausibly be argued, "subjecting the defective child to this sort of institutionalized deprivation amounts to a harm of considerable proportion. In conjunction with the child's purely *medical* disabilities, such *socially* caused burdens may well prompt us to conclude that an early death would be in the best interests of the child. In the absence of better alternatives to institutionalization, such as it is today in many economically deprived areas, these children should be relieved of potentially crushing burdens by means of an early, merciful death."

This is a plausible argument. There is no question in my mind but that the abuse and neglect attributable to a family's inability to cope or to pay for humane care *can* amount to substantial burdens for the child. And so long as burdensomeness to the child remains the crucial criterion involved in the best interest standard, why shouldn't social as well as purely medical burdens count in our overall calculations of best interest? Indeed, if socially induced burdens can make continued life excessively burdensome for a severely retarded child relegated to the back ward, logic would seem to compel us to weigh them as heavily as any other kind of burden. The important thing, it might be said, is not the *source* of the burden, but its *effect* on the child. If parental neglect, social ostracism, and degrading conditions can make life unbearable, why shouldn't they be a part of our moral deliberations?

In spite of its initial plausibility, this line of reasoning runs afoul of the nondiscrimination principle. If it would be in a wealthy child's best interest to receive life-sustaining therapies, then justice demands that a poor child with a similar medical condition and prognosis also receive the same treatment. Like cases should be treated alike.[41]

True, in our previous discussion of medically induced burdens we suggested that certain defects could be so severe as to render an anomalous child significantly different from a normal child for purposes of moral analysis. We said then that such defects could prompt us to deny the premise that the severely defective child was still "like" the normal child in all relevant respects, and thus to deny the conclusion that such children must receive "like" treatment. We argued, finally, that the severely defective child was unlike the normal child precisely in the sense that its best interests would not be served by further treatment or continued existence. Why not repeat the same argument here with regard, this time, to socially related burdens?

The answer to this eminently sensible question is that, whereas an extremely poor medical prognosis and quality of life can make for a significant moral difference between two babies, the same cannot be said of differences in wealth. Wealth is never a morally significant difference between two potential recipients of basic health care. If a certain medical procedure is deemed to be truly beneficial and offered to a wealthy defective child, it must also be offered to the similarly situated poor child. The latter is certainly not responsible for the lack of concern of her parents, or for their inability to afford dignified and humane followup care. To say that the wealthy child should live (because her defects are not so severe as to make life too burdensome), but that the poor child with similar prognosis should die (because of further burdens imposed by her poverty) is to indulge in the rankest kind of discrimination. For purposes of moral analysis, these two patients are significantly identical, and they should therefore receive similar treatment. That is the demand of simple justice—unless we hold that being rich in itself makes the wealthy defective child more deserving of health care than the poor child. An unlikely concession for us to make.

It might be noted in passing here that a policy permitting us to consider socially induced burdens would not only be profoundly unjust; in addition, such a policy would most likely retard efforts to improve the plight of the institutionalized, poor, defective child. Historically, one of the most potent arguments in favor of social reform has rested on the existence of human suffering and deprivation. Could we expect vigorous efforts to reform public institutions harboring anomalous children once the suffering and neglect that characterize such institutions have become publicly recognized as good reasons for killing defective newborns? Although our respective opinions on this matter are bound to be largely speculative, my sympathies lie entirely with Paul Ramsey's response to the suggestion that a child's poverty ought to count in its constellation of publicly recognized burdens. "That's one way," says Ramsey, "to remove every evening the human debris that has accumulated since morning."[42] For the great majority of these handicapped infants, radical social reform is the answer—not "beneficient" euthanasia.

The procedural implication of this argument about justice is that we should attempt to distinguish clearly between "treatment decisions" and decisions regarding the child's ultimate disposition.[43] In other words, worries about the child's future life with a potentially abusive parent or in a hopelessly inadequate state institution should not be allowed to influence treatment decisions. Even though socially induced burdens can join forces with strictly physical deficits to make a life excessively burdensome to its bearer, we must base our treatment decisions solely on the extent of medical defects. To do otherwise—to take social factors into account—is to act unjustly toward the child.

At this point in the argument, we have clearly entered a "moral blind alley"—i.e., a situation so structured that whatever course we take, we end up doing something morally unacceptable.[44] Ordinarily this structure takes the form of a conflict between two disparate moral concerns—namely, doing good and doing justice. Morality, through the principle of beneficence, requires us to increase people's happiness and mitigate their unhappiness. But another moral principle enjoins us to respect persons' rights and to do justice. When these conflicting demands meet head-on in a concrete situation, the stage is set for an ethical dilemma. How can they be reconciled?

As Ronald Dworkin has contended, a concern for rights and justice can function as a "trump card" in moral and political argument with regard to competing utilitarian considerations.[45] When there is a conflict between the demands of utility and the requirements of justice, we usually say that the former must simply yield to the latter. Pointing to the good or bad *consequences* said to flow from the performance or omission of an act is usually said to be an insufficient (or even *irrelevant*) justification if that act or omission happens to be unjust. For example, we say that a citizen's right to speak freely must be respected, even though rather serious consequences (such as public disturbances) may ensue from the exercise of his right. In this sort of case, it is clear what the right thing to do is—we should let the individual speak—and there really is no serious moral dilemma. One moral concern has simply *eclipsed* another.

Sometimes circumstances may be so extreme and the consequences so dreadful that the priority of justice can no longer be maintained. When this happens, an agent enters a moral blind alley. He must choose, but all of his alternatives are morally unacceptable. If he does the "right" thing, the act required by strict justice, then truly catastrophic results will follow; but if he violates the demands of justice, he avoids terrible consequences at the cost of doing a deed that should never be done. In such cases, one may prudently decide to violate rights and justice in order to avoid catastrophe, but this does not always count as a complete *justification*. "Moral traces" of the demands of justice remain, even when circumstances make it plain that justice must be violated. Simply put, it does not become *all right*: we are left with dirty hands.[46]

Winston Churchill faced such a moral dilemma in the thick of World War II—should he launch rocket attacks on German civilian populations, thereby killing hundreds of *innocent* persons, or risk losing the war?—and I suspect that we are faced with a similar dilemma in the case of the impoverished defective newborn. If we take the child's socially caused burdens into account—e.g., the likelihood of being neglected in a state institution—we may rightly decide that death would be better than such a fate. Allowing such a child to die would be a merciful act averting an unbearable life for the child; but it would also be unjust. As we have seen, it is profoundly unjust to afford differential treatment to anomalous newborns on the basis of socioeconomic factors. However, if we avert our eyes from the child's subsequent fate in a barren state institution, we act justly, but possibly consign the child to a miserable life in a warehouse for discarded humans—a catastrophic result, to be sure.

It would appear, then, that we are confronted by a genuine moral dilemma occasioned by social conditions of poverty and inequity. Here, as elsewhere in the field of bioethics, straightforward appeals to notions of right and justice—abstracted from a social context of profound injustice—can lead to morally disastrous results.[47] So long as these unjust social conditions persist, so long as the rich defective child enjoys a bearable quality of life denied to its poor counterpart, the door of the neonatal nursery will continue to issue upon a moral blind alley. Rather than serve as an invitation to resignation, however, this conclusion underscores the necessity of increased social (i.e., governmental) support of the handicapped. Surely, any society that insists upon a controversial strict standard of justice concerning nontreatment decisions—a standard that will weigh heaviest upon those parents most in need of psychological and financial support—surely, such a society *thereby* incurs a corresponding duty to support those who must bear the burden it imposes.

What Standard Will Measure the Burden?

In the previous section we reasoned from the competent adult's conviction that death is not always the worst thing that could befall a person to the conclusion that continued existence need not always be in the best interests of a nondying defective child. Now we must inquire into the nature of the standard according to which burdens to the child are to be measured.

There are essentially two possibilities: we can either adopt the point of view of the rational adult or that of the anomalous child. According to some philosophers, we ought to adopt the viewpoint of the mature adult or the so-called "reasonable person."[48] "Since incompetent creatures are incompetent," writes Joseph Margolis, "we judge as rational adults of normal sensibilities would. It is in this sense that the decision to bring to an end the lives of selected fetuses and infants . . . is *conceptually parasitic* on the decision of the competent to end their own lives. . . ."[49]

Although this approach is flawed, there is a certain appealing modesty to it. It does not require us to plumb the depths of the incompetent person's psyche to discern there what the incompetent patient himself "really wants." Such modesty is becoming when the patient for whom we are to decide has not left us any evidence of how he or she would decide; and it is especially becoming when the patient lacks the very capacity to develop a value system in the first place. Although courts have recently ascribed to themselves or to guardians the right to make so-called "substituted judgments" for voiceless patients, and although the proxy decisionmakers have been legally charged with the (impossible?) task of deciding *as the patient would decide,* were he or she miraculously to become competent for an instant, I suspect that at bottom they end up attempting to judge as any rational adult of normal sensitivities would. And for good reason. In the absence of explicit or implicit value judgments from the incompetent patient, what else do we have to go on? How do we know whether the permanently incompetent patient would judge life under present circumstances to be worth the candle or not? Far from being endowed with the Herculean powers of discernment required by such a task, it would appear that often the best we can do is to ask what the "vast majority" of adults would want, should such a fate befall them. This sort of judgment might still be very difficult to render, but at least we would be familiar with the grounds of our decision. We all presumably have some rough idea of when life would no longer be meaningful or worth living *for us*.

In spite of its commonsensical appeal, two different objections have been lodged against invoking a "reasonable person" standard in the case of defective newborns. The first merely points to the probability of abuse: a reasonable person standard, based on what the vast majority of people would want, leaves the child unprotected against the whims and vagaries of normal adults.[50] This prospect leads some critics to reject out of hand *any* consideration of quality of life in the treatment of newborns.

This objection is itself somewhat ambiguous. It could be taken to mean that a reasonable person standard is theoretically acceptable, but would tend to be too easily abused in practice. This objection may or may not be sound. Empirical studies would have to be adduced showing that the likelihood of abuse is so great that any and all quality of life judgments based on adult sensibilities should be banished. In the absence of such evidence, an abrupt dismissal of a best interests standard would appear to be unsupported.

Another objection goes to the heart of the matter. According to this version, the problem lies in the reasonable person standard itself, not in our penchant for misusing it. Any standard of judgment based on the sensibilities of *normal* adults, it is claimed, will necessarily be prejudicial to the best interests of *defective* children.[51] In contrast to the point about the inevitability of abuse, this objection assumes the appropriateness of a best interest standard, but insists that any criterion based on the values of normal adults is bound to systematically distort our judgments of what is truly in a child's best interests. This version of the objection thus poses a direct challenge to Margolis's assumption that decisions on behalf of incompetent children must be conceptually parasitic on the standards appealed to by normal adults in deciding to end their own lives.

Come to think of it, why *should* the sentiments of normal adults be used as the touchstone of meaningful life for defective newborns? Although most normal adults are no doubt well-meaning and genuinely concerned not to abuse their standard of judgment, they are undoubtedly biased in favor of normalcy. Being normal themselves, competent adults have naturally pitched their value systems on the solid ground of normalcy. But it is precisely through the mediation of these values that adults are supposed to evaluate the defects of the child and assess their implications for the value of continued life. Having grown accustomed to the social and intellectual satisfactions that normalcy makes possible, it is only natural to expect that, were the question put to them, many normal adults would rather die than live without these basic human capacities. But as John Robertson pointed out long ago in a classic treatment of the subject:

> *Yet a standard based on healthy, ordinary development may be entirely inappropriate to this situation. One who has never known the pleasures of mental operation, ambulation, and social interaction surely does not suffer from their loss as much as one who has. While one who has known these capacities may prefer death to a life without them, we have no assurance that the handicapped person, with no point of comparison, would agree. Life, and life alone, whatever its limitations, might be of sufficient worth to him.*[52]

Echoing Robertson's point, the President's Commission recently warned against adopting the viewpoint of normal adults in assessing the best interests of defective children. Although conceding that responsible parties must try to ground their judgments on considerations that would move a *competent* decisionmaker to forego treatment—thereby barring the consideration of idiosyncratic value judgments—the President's Commission strongly recommended that benefits and burdens be weighted from the *incompetent* infant's own perspective.[53] We must ask, not whether a normal adult would rather die than suffer from severe mental and physical impairments, but rather whether this defective child, this child who has never known the satisfactions and aspirations of the normal world, would prefer nothing to what he or she has.

If the best interests of the child is to be our standard of judgment, the viewpoint of the child is clearly the most appropriate measure of benefits and burdens. The standard of the reasonable adult, based as it is on the capacities and expectations of normalcy, would import a systematic bias into our considerations. Everywhere, it would tend to have us see dashed hopes and frustration, rather than opportunities for growth and strategies of meaningful accomodation. Adopting the child's viewpoint would be difficult in practice, but it would certainly conform more closely to the spirit of the best interest standard. The issue, after all, is the welfare of *this child*, not the hopes and fears of adults that might be projected onto the child.

As the Commission rightly points out, this interpretation of the best interest standard yields a very strict policy.[54] For one thing, it would have us fasten onto the question of the child's well-being, to the exclusion of any and all considerations of the child's negative impact on others. Nontreatment of a nondying anomalous child could not be justified on the ground that his ongoing care would drain the family (or the state) of its financial resources or that it would deprive his siblings of the love and attention that they deserve.[55] In addition, this gloss of the best interest standard would clearly have mandated life-saving treatments in most of the cases that have generated public controversy in recent years. Although normal adults might prefer, for themselves, death to mental retardation, it cannot plausibly be argued that the average Down's syndrome child would see death as being in his or her best interest. To the contrary, Down's syndrome persons derive great satisfaction from their lives. From *their* point of view, life is good and worth living, no matter how limited *we* might think their lives to be. The main difference separating them from us lies not in our capacity for enjoying life and for sharing love; rather, it has been plausibly suggested, the main difference might well lie in their "congenital inability to hate."[56]

The Limits of the Best Interests Standard

This standard, interpreted from the anomalous child's viewpoint, is indeed strict—and rightly so. When human lives are at stake, we do not want to substitute, no matter how unconsciously, the fears of normal adults for the genuine best interests of handicapped children. It has not been generally appreciated, however—not even by the President's Commission—just *how strict* this child-relative standard can become when pushed to the extreme limits of human disability.

The first hint of a problem with the application of this standard can be found in Robertson's initial article championing the viewpoint of the defective child. Immediately following his convincing refutation of the claim that death would be a better fate than life for most "mongoloid" and nonambulatory spina bifida children, Robertson undertook to reach the same conclusion in the "worst case"—i.e., "the profoundly retarded, nonambulatory, blind, deaf infant who will spend his few years in the back ward cribs of a state institution. . . ."[57] Even in such an extreme case, where the Commission would no doubt be most likely to concede that "continued existence would not be a net benefit to the infant,"[58] Robertson refuses to sanction nontreatment and death. How do we know, he asks, that such a child with wants, needs, and interests so different from our own would prefer the abyss to his back ward existence? How can we be sure that, for such a child, life and life alone would not be of sufficient worth?[59]

Assuming that the infant is not in pain and has no conscious experience of suffering, Robertson's doubts appear to be well founded. Of course, if this child were also suffering inordinately from the side-effects of invasive therapies or from the underlying condition itself—in addition to his other multiple handicaps—that would be a different story. As we have already seen, unremitting and unrelievable pain can transform a person's continued existence from an unquestioned benefit into a harsh burden. But suppose the child suffers no pain. He lies there in his backward crib blind, deaf, and uncomprehending. Never having experienced the satisfactions of normalcy, the infant is neither horrified by his plight, nor depressed by his neglect in the state institution or by the thought of his impending doom. He just lies there, as if in a vegetative state. Can we say with moral certitude that it is in this child's best interest to die?

I think not. We can say that, for such a severely impaired child, life cannot have much significance or value. Indeed, it is problematical to understand in what sense such a child can have any interests to which a best-interest standard might apply. Nevertheless, it is far from clear that, strictly from this infant's point of view, death would be preferable to such a life. As Robertson implies, the mere act of subsisting on the back ward—vegetating, if you will—might well be deemed "worth the candle" by this child if, miraculously, he could survey his situation with the clarity of a competent person. "It's certainly not much," he might ruefully admit to himself, "but it's better than nothing. They feed me, change me, occasionally cuddle me. And I'm not in any pain to speak of. So why not just play out the meagre hand that I've been dealt?"

This eerie soliloquy, spoken by a child who will never speak, will certainly play to mixed reviews. For some, it will furnish convincing evidence only that the best-interest standard, interpreted from the child's point of view, mandates treatment even in the worst cases (provided, of course, that the child is not in pain). Others, though (correctly) agreeing with this conclusion, will see in it evidence, not that such children should be kept alive, but rather that, in such extreme cases, the best interest standard has been pushed beyond the pale of its capabilities. They will contend, rightly I think, that when pushed this far the best interest standard generates results that conflict with moral common sense.

Consider, for example, the plight of the child born with Trisomy-13. This is a chromosomal disorder associated with severe mental retardation and severe physical impairments. The child's brain is so malformed that it will never possess the capacity to think or to communicate with others. And, as if these deficits were not sufficient, it is doomed to an early death. If treated aggressively and vigilantly, however, the infant might well live for months, or perhaps even for a year. Although probably not properly classified as a "dying patient" according to Surgeon General Koop's taxonomy,[60] this child will soon die, never having had an intelligible thought, never having given or acknowledged love, never having experienced anything beyond simply "being there."

The fact that the child-relative best-interest standard would mandate treatment even in the face of a prognosis bereft of any distinctly *human* potentiality reveals a feature of that standard that has so far gone unnoticed in recent discussions. In such extreme cases, the best-interest standard tends to view the absence of pain as the only morally relevant consideration. No matter whether the infant is doomed to a life of very short duration, no matter that the child lacks the capacity for any distinctively human development or activity; so long as it does not experience any severe burdens, interpreted from its own point of view, the fact that the child can anticipate no distinctly human benefits is of no moral consequence.

But this seems wrong. The presence or absence of such characteristics as the ability to think, to communicate, to give and receive love, seems to be highly relevant from a moral point of view. Indeed, in the absence of these capacities, it is problematical in the extreme how we can attribute any human interests to the child on which a best-interest standard might operate. By narrowing the range of meaningful data to the presence or absence of unrelievable pain, the best interest standard consigns itself in extreme cases to operating in a moral vacuum. The result is an indiscriminate mandate to treat, to keep alive, that flies in the face of common sense. When confronted with such cases, one wants to ask, not whether treatment will further the infant's best interests, but rather whether the child's impoverished level of existence is worth sustaining.

Beyond the Best Interest Standard

Far from advancing the conclusion that even the worst case ought to receive life-sustaining treatments, our hypothetical soliloquy reveals the limits of the best-interest standard; and, in so doing, it underscores our need for a supplementary standard geared to the presence or absence of distinctly human capacities.[61] The ethical principle that justifies this standard is the proposition that biological human life is only a relative good. In the absence of certain distinctly human capacities—e.g., for self-consciousness and relating to other people—the usual connection between biological life and our notion of the good is effectively severed. Just as the presence of unrelievable pain can preclude the attainment of those basic human goods that make life worth living, so the absence of fundamental human capacities can render a life valueless, both to its possessor and to others.[62] Without these qualities, no distinctly human good can be achieved. When this point is reached, the duty to sustain life loses its hold on caregivers.

Since such grievously afflicted children have no distinctly *human* capacities, and thus no *human* interests, the activity of keeping them alive is literally *pointless* from the moral point of view. Indeed, I would go so far as to argue that, in these worst cases, it is a mistake to inquire about the best interests of such children or to insist that decisions to terminate treatment be made solely in terms of the child's good.[63] As our previous discussion of the best-interest standard revealed, the only interest that can intelligibly be attributed to children lacking self-consciousness or relational potential is an interest in avoiding pain and suffering. Since, for them, continued existence does not offer the prospect of achieving any human good, we cannot sensibly attribute to such children an interest in sustaining their lives. Consequently, those who bear the responsibility for them are under no ethical obligation to prolong their lives by means of medical treatments.

In contrast to the misapplied best-interest standard, which sought to fasten onto the *subjective* preferences of the severely defective child, this "relational potential standard" must issue from a social, *intersubjective* inquiry into the conditions of a valuable human life. Although any attempt to define the parameters of "meaningful human life" will doubtless meet with suspicion—or even with extreme hostility—this sort of inquiry makes a great deal more sense than the convoluted speculations required by a best-interest standard pushed beyond its limits. The search for a threshold of meaningful human life might well be difficult and fraught with the danger of abuse—how many pogroms and genocidal campaigns have been justified by their victims' alleged lack of "basic humanity"?—but the search for the secret preferences of patients lacking the capacity for self-knowledge and human relations is, I would argue, an essentially misguided venture.

The futility of this search is illustrated by the paradoxical subjective tests the courts have applied in extreme cases. In the famous Saikewicz case, for example, the judge bids us to reach a decision that "would be made by the incompetent person, if that person were competent, but taking into account the present and future incompetency of the individual as one of the factors which would necessarily enter into the decision-making process of the competent person."[64] Although the language of the President's Commission avoids the transparent absurdity of the *Saikewicz* formula, its recommendation amounts to the same thing: We are to base our judgments on grounds that "would lead a *competent* decisionmaker . . . to forego treatment" while, at the same time, "adopting the viewpoint of the *incompetent* patient."[65] Given the impossibility of this charge in the case of children lacking any relational potential or capacity for choosing a system of values, we are faced with two alternatives. We can either attempt to sustain the lives of such patients in spite of their empty prognoses, or we can engage in the risky business of designating a threshold of meaningful human life. The latter alternative might well be dangerous,[66] but the former is pointless and burdensome to parents and society.

The Beginning of Wisdom

Our reflections on the best interest standard and its limitations have thus revealed two morally relevant distinctions between normal infants and certain severely anomalous children. Just as a child might be excessively burdened by his treatments or underlying disease, so a child may lack certain basic human capacities. I have argued that such pain and lack of capacities mark significant moral differences between these defective children and their normal counterparts. Just as we ought to extend similar treatment to similar cases, we should also be prepared to extend dissimilar treatment to dissimilar cases. Since the nondiscrimination principle forbids differential treatment only in the absence of morally relevant distinctions between people, and since excessive burdens or the lack of relational potential constitute relevant distinctions, the nondiscrimination principle does not forbid nontreatment in the sorts of cases we have been discussing.

As the complexity of the foregoing discussion attests, we are confronted here with an enormously difficult and complex moral problem. Contrary to the opinion of those who would simplify this problem beyond all recognition—either by mandating treatment for all children, or by allowing parents or health professionals unbridled discretion—I have tried, following Kierkegaard's example, ". . . to create [or at least acknowledge] difficulties everywhere."[67] The beginning of wisdom here is to come to grips with this complexity, not to sweep it under the rug. This means acknowledging the pull of fundamental but contradictory moral imperatives—e.g., to sustain life and to ameliorate suffering—and it also means acknowledging the vast spectrum of anomalous conditions and possible interventions. Our public policy must be grounded on respect for this complex reality, not on one or two sensational cases.

The second step on the way to wisdom is the realization that there are several ways to misuse our powerful new technologies. "Pro-life" advocates are right to insist that we can err by withholding life-sustaining treatments that would serve the best interests of defective children. They warn that withholding care on the grounds of the child's best interests or the child's lack of relational potential will lead us down the slippery slope of abuse and neglect. And they conclude that the nondying defective child must *always* be treated. In response, it must be said that we can also err and do harm by imposing excessively burdensome treatments on certain children, or by treating them when it will do no earthly, human good.

The issue is thus shot through with ethical ambiguity. Most anomalous children should be treated, some should be allowed to die. Substantive principles are available, but their application is fraught with difficulty and danger. Although this sort of pervasive ambiguity is difficult to live with, we can be sure that attempts to ignore it, to reduce the problem to a simple formula, will lead to an illusory and counterproductive quest for moral certainty.

Postscript

It seems that hardly a month goes by these contentious days without a significant new announcement regarding the "Baby Doe Question." Since this article was originally drafted, we have seen the case of Baby Jane Doe come and go, leaving behind a legacy of appelate court decisions and intensified public debate.[68] Following the demise of the Infant Doe Rule at the hands of the Second Circuit Court of Appeals, which ruled that the Department of Health and Human Services' (DHHS) Final Rule on Baby Doe lacked legislative authorization in the Rehabilitation Act of 1973, the advocates of increased governmental intervention next turned to the legislative branch.

Since the legal basis of the administration's "anti-discrimination" theory had fallen through, a new platform had to be established in the form of fresh legislation. During the Summer and Fall of 1984, an unlikely coalition of liberal and conservative Senators joined forces with an equally unlikely assortment of "prolife" groups, disability advocates, and the American Academy of Pediatrics to forge a legislative "compromise." The result was a crucial "Baby Doe" clause incorporated in HR 1904, the 1984 amendments to the Child Abuse Prevention and Treatment Act.[69]

Although this clause was hailed by many as a significant ethical and legal compromise between previously hostile and unyielding factions,[70] subsequent proposed regulations from DHHS indicate that the "prolife" ideologues within the Reagan administration are incapable of appreciating the complexity of this intractable problem. In the face of a society equally divided on this excruciatingly difficult moral issue; in the face of repeated court decisions questioning the propriety, not to mention the legality, of imposing a rigid obligation to treat all nondying impaired infants; and, finally, in the face of a hard-won legislative compromise that held out the promise of *some* reasonable discretion for caregivers—in the face of all these countervailing factors, DHHS has drafted a set of guidelines that threaten to take us right back to the substance of the simplistic Infant Doe Rule. "Plus ça change, plus c'est la même chose."

Although this is not the place to develop a full account and critique of the new legislation and proposed regulations, a task I have undertaken elsewhere,[71] a few words might still be in order to establish the near equivalence of the proposed DHHS regulations and the substantive principles animating the old Infant Doe Rule, and thereby to establish the enduring relevance of the present ethical critique.

The Child Abuse Prevention and Treatment Act defines a new category of medical neglect: the withholding of "medically indicated treatment" from "disabled infants with life-threatening conditions."[72] This variant of medical neglect is further defined as "the failure to respond to the infant's life-threatening conditions by providing treatment (including appropriate nutrition, hydration, and medication) that, in the treating physician's or physicians' reasonable medical judgment, will be most likely to be effective in ameliorating or correcting all such conditions. . . ." Importantly, the Act contains a set of enumerated exceptions, which provide that treatment shall *not* be required when: "(A) the infant is chronically and irreversibly comatose; (B) the provision of such treatment would (i) merely prolong dying, (ii) not be effective in ameliorating or correcting all of the infant's life-threatening conditions, or (iii) otherwise be futile in terms of the survival of the infant; or (C) the provision of such treatment would be virtually futile in terms of the survival of the infant and the treatment itself under such circumstances would be inhumane."

Although this standard is described as a "compromise" between medical groups seeking some measure of professional discretion and the "prolife"/disability groups, it can hardly be said to effect a genuine compromise either with the President's Commission's "best interest" principle or with my own supplementary "quality of life" standard. Both of the latter principles are strict, ruling out nontreatment in the case of most Down's syndrome and spina bifida babies, but would nevertheless permit nontreatment in the most severely afflicted nondying infants.[73] But according to the new law, unless a child is deemed to be either "chronically and irreversibly comatose" or engaged in the process of dying, he or she must be treated.

And although the Act *seemed* to accord some measure of reasonable discretion to physicians—more on this semblance later on—it certainly did not offer any sort of compromise to the advocates of parental discretion in difficult and ambiguous cases. Notwithstanding DHHS's rather disingenuous claim that ". . . the parents of the disabled infant also play a crucial role in this process . . .,"[74] the only discretion accorded to parents by this Act is the "discretion" to treat. Although a recent Gallup Poll indicated that 43% of the respondents would ask their doctor not to keep a severely handicapped or deformed infant alive,[75] the Act regards parental participation in such decisions to be tantamount to child abuse.

In spite of the Act's shortcomings, many observers initially welcomed it as a limited but promising advance over the legal standard set by the defunct Infant Doe Rule. Neonatologists interpreted the vague language of the Act—language that spoke of the "process" of dying, and of "virtually futile" and "inhumane" treatments—as allowing them to make, not only "reasonable medical judgments," but also a limited number of reasonable *ethical* judgments based implicitly on a "best interests" or constrained "quality of life" principle. As one prominent neonatologist put it, "It's not perfect, but we can certainly live with it."[76]

The recent publication of DHHS's "proposed rules" put an abrupt end to these optimistic predictions. DHHS has identified and rooted out every instance of vague and flexible language in the Act, substituting in every case its narrow and rigid interpretation of what is "medically indicated." The proposed regulations explicitly require that treatment decisions are not to be based on "subjective quality of life" concepts.[77] They specify that the statutory exception for treatments deemed merely to "prolong dying" applies only when death is *imminent*. Likewise, they interpret the statutory exception for treatments that would not ameliorate or correct all of the infant's life-threatening conditions as applying only when each condition is *imminently* life-threatening.[78] The proposed regulations would thus appear to require corrective surgery on the esophageal atresia of a Trisomy-13 or -18 infant, despite the severity of his underlying defect and his predictably brief life—a requirement that flies in the face of current medical practice and moral common sense.

In brief, the proposed regulations, should they be made final, promise to transform a remarkable effort at moderation and compromise into a mere legislative retread of the Infant Doe Rule. Fortunately for the analysis of this essay, but unfortunately for many extraordinarily burdened newborns and their parents, the foregoing ethical critique of the Infant Doe Rule remains sadly relevant to the administration's latest initiative in legislating a "prolife" morality.

Acknowledgments

I am indebted to Nancy Rhoden, Alan Fleischman, Nancy Dubler, Nick Rango, Betty Wolder Levin, and Liz Emrey for their trenchant criticisms of an earlier draft of this paper. Portions of this article have been previously published in the *Milbank Memorial Fund Quarterly* (Winter 1985) and in the *Hastings Center Report* (April 1984), and are reprinted here with the kind permission of the publishers.

Notes and References

[1] A Gallup survey taken in May 1983 showed the nation to be evenly divided on the question whether severely handicapped or deformed newborns should be kept alive. Forty three percent of the respondents said they would ask their doctor not to keep such a child alive. Forty percent took the opposite position. *Richmond Times Dispatch* (June 2, 1983).

[2] The most influential and articulate champions of parental discretion are Raymond Duff of the Yale Medical School and Joseph Goldstein (and colleagues), also of Yale University. See Raymond S. Duff and A. G. M. Campbell, "Moral and Ethical Dilemmas in the Special-Care Nursery," *New England Journal of Medicine* **289**, no. 17 (October 25, 1973), pp. 890–894; and J. Goldstein, A. Freud, and A. J. Solnit, *Before the Best Interest of the Child* (New York: The Free Press, 1979).

[3] Goldstein et al. contend that parents should have discretion to withhold life-sustaining care from their child if he or she is unlikely to ". . . have either a life worth living or a life of relatively normal growth. . . ." So long as reasonable people can disagree about a particular

case, and so long as the parents want the child to die, Goldstein et al. would allow them to withhold treatment—even if reasonable people would conclude that the child's best interests would be served by continued life. See *Before the Best Interest of the Child.* According to Dr. Duff, ". . . responsible decision makers cannot avoid some "tragic choices"—that is, at times knowingly sacrificing, perhaps unfairly, one person's good or life in order to protect another's." Raymond S. Duff, "Counseling Families and Deciding Care of Severely Defective Children," *Pediatrics* **67,** no. 3 (March 1981), p. 316.

[4]John Rawls, *A Theory of Justice* (Cambridge: Harvard University Press, 1971), p. 86.

[5]Ronald Dworkin, *Taking Rights Seriously* (Cambridge: Harvard University Press, 1977), pp. 31–39.

[6]Some philosophers have argued that the defective newborn controversy ought to be viewed as a logical extension of the abortion issue. If we presently regard defective *fetuses* both as prime candidates for abortion and as being interchangeable with (future) normal fetuses, they argue, why not regard defective *newborns* in the same way? The crucial assumption of this argument is that there is no significant moral difference between a second or third trimester fetus and a neonate. See Peter Singer, *Practical Ethics* (Cambridge: Cambridge University Press, 1979), pp. 131–138.

[7]The terminology is Paul Ramsey's. The following discussion owes much to Ramsey's penetrating analysis of the ordinary/extraordinary distinction in his fine book, *Ethics at the Edges of Life* (New Haven: Yale University Press, 1978), pp. 145–160.

[8]Ibid., p. 270.

[9]In addition to technological complexity and statistical ordinariness, a treatment's degree of "invasiveness" is often mentioned as a measure of "extraordinary" means. Although this interpretation is alert to the consequences of a therapy for patients—i.e., whether or not it is "invasive"—it remains a categorical approach just the same, and thus remains vulnerable to the same criticism that we have been developing here. Whether or not a treatment is deemed "invasive"— whatever that means—appears irrelevant to the issue of whether it is *morally* obligatory. Depending on the circumstances, highly invasive heart surgery may be morally required, whereas such non-invasive measures as an intravenous drip may be considered optional. In at least one jurisdiction, invasiveness has been (unwisely) enshrined as an element of a legal test determining which treatments may be foregone. See *In Re Quinlan,* 70 NJ 10 (1976): "We think that the State's interest *contra* weakens and the individual's right to privacy grows as the degree of bodily invasion increases and the prognosis dims" 355 A. 2d at 664.

[10]Gerald Kelly, SJ, *Medico-Moral Problems* (St. Louis: Catholic Hospital Association, 1958), p. 129. Quoted in Paul Ramsey, *The Patient as Person* (New Haven: Yale University Press, 1970), p. 122.

[11]Ibid., pp. 113–164.

[12]"[S]ome people prefer to speak of 'proportionate' and 'disproportionate' means. In any case, it will be possible to make a correct judgment as to the means by studying the type of treatment to be used, its degree of complexity or risk, its cost and the possibilities of using it, and comparing these elements with the result that can be expected, taking into account of the state of the sick person and his or her physical and moral resources." Sacred Congregation for the Doctrine of the Faith, *Declaration on Euthanasia* (Vatican City: May 5, 1980). Reprinted in President's Commission for the Study of Ethical Problems in Medicine and Biomedical and Behavioral Research, *Deciding to Forego Life-Sustaining Treatments* (US Government Printing Office, March 1983), pp. 300–307.

[13]See James Rachels, "More Impertinent Distinctions," in Thomas A. Mappes and Jane S. Zembaty, eds., *Biomedical Ethics* (New York: McGraw-Hill, 1981), pp. 355–359.

[14]Ramsey, *The Patient as Person*, pp. 132–136.

[15]According to Dr. Koop, "For such [dying] infants, neither medicine nor law can be of any help. And neither medicine nor law should prolong these infants' process of dying." Statement before Hearing on Handicapped Newborns, Subcommittee on Select Education Committee on Education and Labor, US House of Representatives (September 16, 1982). Quoted in President's Commission, *Deciding to Forego Life-Sustaining Treatment*, pp. 219–220 n. 81.

[16]"The chief problem of the dying is how not to die alone. To care, if only to care, for the dying is, therefore, a medical–moral imperative. . . ." *The Patient as Person*, p. 134.

[17]Ramsey, *Ethics at the Edges of Life*, pp. 191 ff.

[18]Ramsey, *The Patient as Person*, p. 187.

[19]Department of Health and Human Services, "45 CFR Part 84: Nondiscrimination on the Basis of Handicap; Procedures and Guidelines Relating to Health Care for Handicapped Infants; Final Rule," *Federal Register* **49**, no. 8 (Thursday, January 12, 1984), pp. 1622–1654. For a brief discussion of subsequent Federal standards enacted as this article goes to press, see the postscript on p. 125 below.

[20]"Section 504 does not compel medical personnel to attempt to perform impossible or futile acts or therapies. Thus, [it] does not require the imposition of futile therapies which merely temporarily prolong the process of dying of an infant born terminally ill." *Federal Register* **48**, no. 129 (Tuesday, July 5, 1983), p. 30852.

[21]Ramsey, *Ethics at the Edges of Life*, pp. 154, 181–188.

[22]"Section 504 is in essence an equal treatment, non-discrimination standard." *Federal Register* **48**, no. 129, at p. 30848.

[23]This legal term refers to groups that have been subjected to discrimination and lack the power to achieve redress of their grievances through ordinary political channels.

[24]*In re Infant Doe*, No. GU 8204-00 (Cir. Ct. Monroe County, Indiana, April 12, 1982), writ of mandamus discussed *sub nom. State rel. Infant Doe v. Baker*, No. 482 S 140 (Indiana Supreme Court, May 27, 1982). See Richard A. McCormick, SJ and John J. Paris, SJ, "Infant Doe: Infanticide or an Acceptable Medical Option?" *America* 313 (April 23, 1983). For the angry response of the father of a Down's syndrome child, see George F. Will, "Starving a Newborn to Death," *New York Daily News* (Sunday, April 25, 1982), p. 47.

[25]This chronology is taken from George Annas, "Disconnecting the Baby Doe Hotline," *The Hastings Center Report* 13 (June 1983), pp. 14–16.

[26]For an interesting survey of physicians' and nurses' likely responses to these and related questions, see Betty Wolder Levin's essay in the appendix of this volume.

[27]See note 22, *supra*.

[28]Notwithstanding the administration's persistent claim that its Infant Doe Rule merely reinforces existing nondiscrimination law, two federal courts have rejected the theory that nontreatment of defective newborns constitutes discrimination as understood by the framers of Section 504. See the opinion of District Judge Wexler in the case of Baby Jane Doe, *US v. University Hospital, State University of New York at Stony Brook*, United States District Court, Eastern District of New York, November 17, 1983; see also the opinion of Judge Pratt on appeal: "Our review of the legislative history has shown that congress never contemplated that section 504 of the Rehabilitation Act would apply to treatment decisions involving defective newborn infants when the statute was enacted in 1973, when it was amended in 1974, or at any subsequent time." *US v. University Hospital, State University of New York at Stony Brook*, US Court of Appeals, 2nd Circuit, Docket No. 83-6343 (February 23, 1984).

[29]48 *Federal Register* at 9631, amending 45 CFR ¶84.61.

[30]According to President Reagan's interpretation of Section 504, "That law forbids recipients of federal funds from withholding from handicapped citizens, simply because they are handicapped, any benefit or service that would ordinarily be provided to persons without handicaps." Quoted in final Infant Doe Rule, *Federal Register* **48,** (January 12, 1984), p. 1622.

[31]*Ibid.*, p. 1622.

[32]Alan H. Goldman, *Justice and Reverse Discrimination* (Princeton: Princeton University Press, 1979), p. 23.

[33]President's Commission for the Study of Ethical Problems in Medicine and Biomedical and Behavioral Research, *Deciding to Forego Life-Sustaining Treatments* (US Government Printing Office, March 1983), pp. 197–229.

[34]Versions of this argument appear in Ramsey, *Ethics at the Edges of Life* (New Haven: Yale University Press, 1978), pp. 206–207, 240 n. 11; and in Paul Camenisch, "Abortion for the Sake of the Fetus," *The Hastings*

Center Report **6** (April 1976), pp. 38–41. See also *Gleitman v. Cosgrove*,
   49 NJ 22, 227 A.2d 689 (1967): "It is basic to the human condition to
   seek life and hold on to it however heavily burdened."

[35]Philippa Foot, "Euthanasia," *Philosophy and Public Affairs* **6** (1977), pp.
   94–96.

[36]Ramsey, *Ethics at the Edges of Life*, pp. 154–59, 225.

[37]President's Commission, *Deciding to Forego Life-Sustaining Treatment*, p.
   218. On closer inspection, this particular gloss of the best interest
   standard turns out to be dangerously overinclusive. Continued existence is not a "net benefit" for many people who would nevertheless
   opt for life over death. Here we should be talking about handicaps so
   severe that continued existence would be marked by *constant pain and
   suffering*. This lapse in the Commission report was called to my
   attention by Alan Fleischman.

[38]See, e.g., Norman Fost, "How Decisions are Made: A Physician's View,"
   in C. A. Swinyard, ed., *Decision Making and the Defective Newborn*
   (Springfield, IL: Charles C. Thomas, 1978), pp. 220–230.

[39]This problem is compounded by our ability to diminish a child's eventual
   degree of disability by means of vigorous medical and educational
   interventions.

[40]For a good description of how bad the conditions can get in such institutions, see *New York State Association of Retarded Children v. Rockefeller* 357 Fed. Supp. 752 (Eastern Division, NY 1973).

[41]The same can be said on behalf of children born to parents, rich and poor
   alike, who are unable or unwilling to love and nurture them. The
   parents' inadequacies are not the fault of their impaired children.

[42]*Ethics at the Edges of Life*, p. 203.

[43]Ibid., p. 202.

[44]Thomas Nagel, "War and Massacre," in Marshall Cohen et al., eds., *War
   and Moral Responsibility* (Princeton: Princeton University Press, 1974),
   p. 23.

[45]Dworkin, *Taking Rights Seriously*, p. xi, 171–172.

[46]Nagel, "War and Massacre," pp. 16–17. See also Robert Nozick, "Moral
   Complications and Moral Structures," *Natural Law Forum* **13** (1968),
   pp. 1–50.

[47]This theme is developed at length in my essay, "The Right to Die on the
   Slippery Slope, *Social Theory and Practice* **8** (Fall 1982), pp. 285–328.
   See also Nancy Rhoden, "The Limits of Liberty: Deinstitutionaliza-
   tion, Homelessness, and Libertarian Theory," *Emory Law Journal* **31**
   (Spring 1982), pp. 375–440.

[48]See Robert Veatch, *Death, Dying, and the Biological Revolution* (New Haven:
   Yale University Press, 1976), pp. 124–136; and Joseph Margolis,
   "Human Life: Its Worth and Bringing It to an End," in Marvin Kohl,
   ed., *Infanticide and the Value of Life* (Buffalo, NY: Prometheus, 1978),
   pp. 180–191.

[49]In Kohl, *Infanticide and the Value of Life,* p. 190.

[50]Perhaps the framers of the Infant Doe Rule had this sort of argument in mind when they decided to characterize all quality of life judgments as being "subjective." They may well have reasoned that since reasonable people can and do disagree over whether a certain kind of life is "worth living," such judgments must express nothing more than mere subjective *preferences* and cannot, therefore, provide sufficient support for a firm public rule. See *Federal Register* **48** (July 5, 1983), p. 30847.

[51]John Robertson, "Involuntary Euthanasia of Defective Newborns," *Stanford Law Review* **27** (1975), p. 254.

[52]Ibid.

[53]*Deciding to Forego Life-Sustaining Treatments,* p. 219.

[54]Ibid.

[55]For an opposing view, see Carson Strong, "Defective Infants and their Impact on Families: Ethical and Legal Considerations," *Law, Medicine and Health Care* (September 1983), pp. 168–181.

[56]Louis Lasagna, "Murder Most Foul," *Reader's Guide: The Sciences* **22,** no. 6 (August–September 1982). Dr. Lasagna is the father of a twenty-year-old with Down's syndrome.

[57]Robertson, "Involuntary Euthanasia of Defective Newborns," p. 254.

[58]*Deciding to Forego Life-Sustaining Treatments,* p. 218.

[59]Robertson, "Involuntary Euthanasia of Defective Newborns," p. 254.

[60]According to Dr. Koop, "For such [dying] infants, neither medicine nor law can be of any help. And neither medicine nor law should prolong these infants' process of dying." Statement before Hearing on Handicapped Newborns, Subcommittee on Select Education Committee on Education and Labor, US House of Representatives (September 16, 1982). Quoted in President's Commission, *Deciding to Forego Life-Sustaining Treatment,* pp. 219–220 n. 81.

[61]Such a standard is worked out by Richard McCormick, SJ in "To Save or Let Die: The Dilemma of Modern Medicine," *Journal of the American Medical Association* **229,** no. 2 (July 8, 1974), pp. 172–176. See also Nancy K. Rhoden and John D. Arras, "Withholding Treatment from Baby Doe: From Discrimination to Child Abuse," *Milbank Memorial Fund Quarterly* (Winter, 1985), pp.

[62]This is not to deny the fact that families can derive great satisfactions from caring for such severely impaired children and may well desire to keep them alive. It remains true, nevertheless, that this particular reason for sustaining their lives has nothing to do with the *child's* best interests.

[63]Even McCormick confuses his standard based on lack of basic human capacities with the best-interest standard. See "To Save or Let Die," p. 176. In a more recent article, coauthored by John Paris, SJ, McCormick compounds this confusion by arguing that the capacity for

human relationships should be viewed as a "summary" of the burden–benefit evaluation. The conflation of these standards might well
be appropriate in cases where a child's pain and suffering is the
source of its inability to relate to others, but not when its lack of
capacity for personal relations results from lack of awareness unac-
companied by pain. See "Infant Doe: Infanticide or an Acceptable
Medical Option?" *America* (April 23, 1983).

[64]*Superintendent of Belchertown State School v. Saikewicz*, 370 N.E. 2d 417
(1977).

[65]*Deciding to Forego Life-Sustaining Treatment*, p. 219.

[66]Whether such criteria should be made explicit and serve as the basis for
what Robert Burt calls a "generalized community validation of the
physician's death-dispensing role" is a very difficult question requiring sustained thought and discussion. Physicians and many others
are calling for increased clarity in these matters, no doubt because of
their (justified) fear of legal liability; on the other hand, critics like
Burt argue that the establishment of *public* criteria of "meaningful
life" will perforce have a certain "bloodless" quality about it—a
bloodlessness "which begins to obscure the fact that we are engaged
in a very bloody business." Burt concludes that the least worst
alternative would be to leave restrictive laws in place while accept-
ing, if not applauding, the continued regime of many ignored law
violations by parents and physicians. Whether this sort of double
standard could be sustained in today's highly politicized climate is,
of course, a debatable question. See Robert A. Burt, "Authorizing
Death for Anomalous Newborns," in Aubrey Milunsky and George
Annas, eds., *Genetics and the Law* (New York: Plenum, 1976), pp.
435–450.

[67]Robert Bretall, ed., *A Kierkegaard Anthology* (New York: Modern Library,
1936), p. 194.

[68]See citations in note 28, supra.

[69]HR Conference Report No. 98-1038, 98th Congress, 2nd Session, pp. 19ff.

[70]In brief, the controversy pitted representatives of organized medicine,
including the American Academy of Pediatrics, against an im-
pressive array of "prolife" and "disability rights" organizations, such
as the Association for Retarded Citizens and the Down's Syndrome
Conference. Not everyone joined the fold. The American Medical
Association refused to endorse the Act, apparently on the ground
that it amounted to unwarranted Federal intrusion into the practice
of medicine; whereas the American Life Lobby withdrew at the last
moment in protest over (what it perceived as) the Act's dangerously
vague language.

[71]Nancy Rhoden and John Arras, "Withholding Treatment from Baby Doe:
From Discrimination to Child Abuse," *Milbank Memorial Fund Quarterly* (Winter, 1985).

[72]*Congressional Record—Senate*, June 29, 1984, p. 8951.

[73]For example, a child born with massive hydrocephalus (head 75 cm in circumference, compared to an average 35 cm); extreme microcephaly (indicating minimal brain function); permanent blindness; a high level and extremely large spina bifida lesion, indicating complete and permanent paralysis of the entire lower body, including the urinary and excretory functions, and requiring painful skin grafts. The child's neurologist and pediatrician concur that the child will be crib-bound and almost certainly unable even to recognize her caretakers when she grows up.

[74]"Proposed Rules: Child Abuse and Neglect Prevention and Treatment Program," *Federal Register*, vol. **49,** No. 238 (December 10, 1984), p. 48163.

[75]See note 1, supra.

[76]Interview with Alan Fleischman, MD, director, Division of Neonatology, Albert Einstein College of Medicine-Montefiore Medical Center, Bronx, New York.

[77]Proposed Rules, p. 41863.

[78]Ibid., p. 48164.

The Right to Privacy and the Right to Refuse Care for the Imperiled Newborn

Nancy N. Dubler

Introduction

Decisions about the care of imperiled newborns traditionally occurred in the privacy of homes or the comfort of doctors' offices. Today, these options are more likely to be framed within the confines of major tertiary care medical centers. Thus some brief comment on the character of these institutions and the legal climate that surrounds their function is prerequisite to a discussion of the substantive issue.

Academic medical centers are peculiar institutions. They are, on the one hand, centers for learning and research. They provide opportunities for undergraduate and postgraduate medical education. They are charged with the care of patients. As if such roles were not sufficiently complex and often competitive, these megalithic enterprises exhibit an additional persona: they are corporate entities. Whether they are profit or not-for-profit—voluntary or proprietary—they are established for the purpose of providing care in a financially careful and fiscally responsible manner. The corporate sector of an academic medical center is focused on assuring adequate business function; it must perpetually consider issues of potential liability. With the growth of malpractice litigation, these considerations are obviously substantial.

The law, which at best involves a search for principles of justice and a focus on fairness and equity, may also be used as a tool for intimidation by special minoritarian interests. This use or abuse of legal process must be recognized and opposed whenever possible. Law both helps in the formulation of, and is affected by, developing societal norms. Justice Holmes has commented that case law is a response to the "unconscious preferences of society."[1] In a post-Freudian era we should not be astonished that areas of conscious and unconscious conflict should appear as legal controversy; this controversy operates not only to hone discussion but to intimidate open exploration and bludgeon dissent.

The complex intellectual, academic, and fiscal nature of the medical center may sometimes be a barrier to intellectual exploration. Use of the law to attempt to stifle discussion by single-purpose groups creates a chilling climate for debate and may diminish protections for vulnerable individuals and institutions or for unpopular positions. Confrontation with hard moral and legal dilemmas need not be courted, but it may not, despite these factors, be avoided.

This comment will, in brief, address some issues having to do with the care of children and incompetent persons in general. It will then explore the right to privacy as it developed as the constitutional basis for empowering third parties to refuse care for one who is incompetent. Finally, it will pose three questions: First, can the right to privacy be extended to support withdrawing or withholding care from defective newborns? Second, if so, what sorts of substantive criteria or guidelines exist to guide these decisions? Third, under what sorts of procedures would these decisions be appropriate?

I will offer a qualified yes to question number one and suggest approaches to structure the development of answers to questions number two and three.

Incompetence, Children, and the Law

The law has a long history of affording special protections for those who cannot adequately define and protect their own interests. From 13th Century England, with the enactment of the statute Depraerogativa Regis,[2] the right and the obligation of the king to protect certain classes of incompetent persons was established. The king was responsible under this statute for the protection of idiots, that is, those born without reason, and lunatics, those who later became of diminished or extinguished mental capacity. Even in this early enactment of protections for the incompetent, however, there was unresolved tension.

The kingly statute was designed in part to protect individual persons and in part to protect interests of the state. Although the state recognized some inherent obligation to protect less fortunate persons, it also recognized that incompetent persons who were "landed" (titled with property) had little ability to control the passage of that title. Lands without clear ownership were a primary reason for citizen unrest. Thus it was in the interest of the state to protect the individual, but also to protect its own orderly rule. This intertwining of protections for the benefit of the individual and for the benefit of the state infuse the original state focus on incompetents. Notwithstanding, the state recognized that it is inherently unseemly to permit those without, before or beyond reason to be misused or abused by others.

This concept of the protection of incompetents, known in the law as the doctrine of *parens patriae,* was recognized as a vague obligation by 19th Century England and America. With the poor laws of the mid-nineteenth century in England, the concept of protection—always double-edged to the benefit of the individual and the state—became more finely honed.

In the late nineteenth century, the concept (i.e., *parens patriae*) took on new garb and substantive and procedural specificity in regard to children. In 1899 Illinois passed the first statute establishing a family court specifically empowered to protect children, those of categorically undeveloped competence.[3] The family court in Illinois, and those judicial progeny that followed it in all other states, established a forum to define the substance and execute the process of protecting children.

A schema of three legal categories was established by the original statute and was followed by all other states. This model sets forth three sorts of children in whom the state has a special interest and for whom it has special obligations of protection.

First, there are those children who are juvenile delinquents, that is, who commit acts that for an adult would be a crime. This category assumes that persons who are not yet fully formed and responsible may not be able to conjure up the intent requisite to perpetrate a crime. Moreover, it protects a child and childhood from the societal label of criminal. It requires that misdeeds be dealt with privately and informally without the stigma and perfidy of conviction, sentence, and incarceration.

The second category of children created was that of "status offender"—children who are charged with no criminal activity, but who are classified as "unruly" or "incorrigible" or "persons-in-need-of-supervision." These children behave in ways that their parents or society deem inappropriate for children; they hang out on street corners or refuse to attend school, or refuse to obey their parents. These are not crimes repugnant to and interdicted by the norms of society.

The last category established by the family court was the "neglected child,"—a child who is not provided with the necessities of life, i.e., care, food, shelter. All states established that the inability or refusal of parents to provide "necessary medical care" would be a category of neglect that could be adjudicated and remedied by appropriate court action.

The Right to Privacy

In the last decade various Federal courts and the United States Supreme Court have struggled to define the rights of persons to control their own bodies. The constitutional right to privacy developed as a justification for the right to use contraceptives without state interference.[4] It then became the prime constitutional pinion for the right to an abortion.[5] At the same time it was used to permit competent persons to refuse suggested or life-saving medical care and treatment. Finally it was held to be a major support for the decisions of third parties, whether guardians or courts, to withhold or withdraw life supports from incompetent persons.[6]

Cases extending the constitutional right to privacy to incompetent persons share, with one exception, a fatal flaw. They all speak the language of individual decision, whereas in actuality they depend for their analysis and conclusion on flights of fancy and fiction. This is understandable. The right to refuse treatment and to choose death, if a competent person has not explicitly chosen this course, is a "right" of such fanciful speculation as to foil real analysis and require the creation of myth.

The cases establishing that the right to privacy protects the right of incompetent persons to have others choose death on their behalf proceed in the following manner: They establish that a right to privacy exists. They state that this right protects the right of a competent person to "choose death," if he or she so desires, over the pain, suffering, and risks of invasive and debilitating medical care and treatment. They assert that merely because one is incompetent one does not lose all rights. They properly insist that one does not forfeit constitutional protections by virtue of diminished mental ability. From these cases, we conclude, therefore, that the right to choose death is protected by the right to privacy and may be exercised by another on behalf of an individual unable to choose.

Thus, the "talk" in these cases is about individual rights, decision-making abilities and the extension of constitutional protections for privacy. An analysis of the facts and reasoning in these cases, however, presents a quite different picture. Consider, for example, two of the key cases, those of Karen Quinlan[7] and Mr. Saikewicz.[8] In the *Quinlan* case, Karen was, at the time of the first hearing, in a "permanent vegetative state." There was some minimal and vague evidence introduced at that time relating to her previously expressed preference in regard to unending coma versus death. The court concluded, quite properly, that these statements had been uttered in such a speculative vein, with so little expectation of application, that they were in fact most uncompelling evidence. Thus, the court concluded that it had no really reliable evidence as to Karen's wishes. What it did, absent an adequate prior statement, was to propose an abstract calculus that it then supported by the introduction of an actual fiction. The court stated, "We think that the state's interests contra weakens and the individual's right to privacy grows as the degree of bodily invasion increases and the prognosis dims. Ultimately, there comes a point at which the individual's rights (i.e., to hypothetically choose death) overcomes the state's interest (in preserving life)."[9] Whereas this theoretical statement is couched in the right of privacy—which can be extended to and afforded to incompetents by certain third parties—the factors balanced are those objective criteria of prognosis, benefit, and the degree of bodily invasion.

The court then engaged in a flight of fancy. It used the fictional device of prospective narrative focus to query what would happen if Karen were "miraculously lucid"[10] and could choose death, in order to import a putative decision as a "valuable incident of her right to privacy."[11] The court stated: "We have no doubt, in these unhappy circumstances, that if Karen were herself miraculously lucid for an interval (not altering the existing prognosis of the condition to which she would soon return) and perceptive of her irreversible condition, she could effectively decide upon discontinuance of the life-support apparatus, even if it meant the prospect of natural death."[12] The number of subjunctives in the syntax should give one pause as to the adequacy of the analysis. There is far too much speculation and projection in this paragraph to permit connecting this variation of the right to privacy with the origins of that right in concepts of self-determination, autonomy, and individual choice about matters affecting one's body.

Consider how far the court's flight of fancy has moved from the articulation of the right to privacy found in the early contraception decisions. In those decisions the issue was pre-eminently one of the individual choice; the issue was whether to "bear or beget a child."[13] Despite the arguments that passions are antirational, the right to privacy developed as a protection for actual decision-making abilities of persons in quite specific situations of personal preference. There was no fictional importation of wants, preferences and desires.

The New Jersey Supreme Court in the *Quinlan* case established objective criteria and procedural safeguards to guard the interests of incompetents. It required the convening of a Prognosis or Ethics Committee to review the prognosis and, moreover, to protect medical caregivers and family from the danger of liability for solo action. It was concerned that medical technology not be applied mechanically, against reason, and without recourse. For lack of a ready legal alternative, it forced these proper state concerns into the improper vehicle of the "right to privacy": But choice is not at issue; protection and process are.

As a second example of fiction in legal analysis, consider the case of Mr. Saikewicz, a profoundly retarded 67-year-old man with a mental age of two years, eight months, who was stricken with acute myoblastic monocytic leukemia. The case raised the question of who could refuse chemotherapy for Mr. Saikewicz and under what standards. Testimony established that chemotherapy might extend his life, but would cause him great discomfort that he could neither understand nor ably incorporate into his life's routine. Again, the court talked about the concept that competent and incompetent patients have equal rights. It then asked how these rights could be extended beyond competence; it discussed the right to refuse and the right to privacy of incompetent patients.

The *Saikewicz* court, however, created another fiction. It used the concept of "substituted judgment," which was originally developed in the law of trusts and land transfer to permit delegation of powers.[14] The court here required itself to "don the mental mantle of the incompetent,"[15] that is to ask, "what would this person want if he or she could say?" In regard to Mr. Saikewicz, even more so than in respect to Karen Quinlan, this is a ludicrous question. Karen Quinlan clearly never expressed a valid preference; Mr. Saikewicz never could have expressed such a preference. The language is couched in terms of individual right, but the calculus used requires the assessment of specific objective criteria. The court required the ascertainment of the incompetent's "actual interests and preferences."[16] Since we can dismiss the latter as in no way intelligible, we are left with the question as to Mr. Saikewicz's actual interests. Because he cannot participate in discussions,

these clearly must be determined by others. The specific conclusion offered by the lower court is that "the negative factors
of treatment exceed the benefits."[17] This, I would argue, is
another objective calculus responsive to generally accepted
societal norms. Again, as with the *Quinlan* case, specific procedural safeguards were created. Massachusetts thus required initial reference to the courts to determine the interest
of the incompetent and the proper course of treatment.

These cases attempt to invoke concepts of substituted
judgment and the right of privacy. What they actually do
however is quite different. They struggle to define a set of
calculations according to certain formulas that ask: what
would reasonable people agree is the "right" or "proper"
action for this person, given the abilities, prognosis, risks,
benefits and suffering?

There is an exception to this line of cases about incompetent persons—the case of *Eichner* or *Brother Fox* in New
York State.[18] That case is *in fact* loyal to the concept of the right
to privacy and to the aura of individual choice and self-
determination that originally supported its development. It
permits the withdrawing or withholding of treatment from an
individual only with the prior explicit statement of that person. It then refuses to extend that right to incompetent persons. It rejects the use of the right to privacy; it can find no
acceptable alternative legal theory to permit such decisions.
This case is arguably and properly loyal to the concept of the
right to privacy, but obdurate in its stubborn refusal to struggle with an alternative legal principle. We cannot as a society
want a rule of law that condemns unfortunate incompetents
to treatment at all costs.

The right to privacy as a legal peg on which to hang
difficult medical decisions and their related ethical dilemmas
does not work well with incompetent people who have left no
directive. Yet, almost all agree that incompetent persons must
have someone, under some conditions, who can oppose the
inexorable grinding of the medical machine. Is there a presently existing legal theory that can permit a third party to
refuse? There is certainly no general agreement on an alternative analysis. Thus, the right of privacy has been bent and

battered into supporting the concept that incompetent people must be protected from the rigid and mechanical imposition of medical care. Even if it is not a clear, bright legal theory of rights it has been so applied and is now accepted, at least in many states, as the basis for withholding and withdrawing treatment from adults. Therefore until an alternative is uncovered by the courts, it is acceptable to apply this principle to neonates. This continues the muddle but at least puts neonates at no greater risk than incompetent adults.

Questions

To answer the question posed at the beginning of this comment—can the right to privacy be extended to support withdrawing or withholding care from newborns—I propose that it can be applied as adequately to newborns as to those others for whom it has been used as a support. To do otherwise is to concede decisions in individual cases to the realm of technological feasibility and mechanical capability rather than to the individual needs of the human condition.

We can discuss the right to privacy for neonates in the light of previous cases. We must then recognize that, despite the various approaches taken, the cases have focused on specific objective criteria: intrusiveness, prognosis, suffering, and benefit. The cases uniformly mandate that a process be developed to balance individual cost and benefit.

Consider the second question—how should the theory be applied, and according to what criteria? I would suggest that the next task of neonatology is to develop those criteria —to translate the categories of law into the facts and stuff of neonatal practice. The law not only permits, I have tried to show, but rather mandates this sort of struggle with individual benefit and suffering.

The third question is: who should apply these standards? In partial answer to that, I must first comment briefly on the issue of parental rights. Part of the confusion in the area of neonatology in regard to decision-making arises, I believe, from the improper application of legal doctrines developed to support relationships between parents and developing children; these issues are separable from decisions relating to the life and death of the newborn infant. Parental rights that nurture and direct the growth of children in the context of a family are quite different from the sort of anguished and emotional conflicts that surround the life and death decisions in a neonatal unit.

In the case of *Bellotti v. Baird*,[19] which assessed the rights of adolescents to an abortion with or without parental notice and control, the court offered the following comment: "Indeed, constitutional interpretation has consistently recognized that the parents' claim to authority in their own household to direct the rearing of their children is basic in the structure of society."[20] In education, care, the importation of values, and the development of a moral foundation, parents are paramount. There is a radical disjunction, however, between that sort of parental authority and the ability of the state to grant to parents the decision over the life or death of a newborn.

Therefore, in answer to questions about who should decide, let me first propose who should not. These are not the sorts of decisions that the law should grant exclusively to parents, subject as they must be to wildly conflicting emotions and loyalties. Nor are these necessarily decisions for courts, given the public nature of their process, and the emergent demands of these decisions. The court process is cumbersome, expensive, public, and staffed by persons with no particular interest or expertise in these issues. Cases demonstrate that judges may be as likely, despite legal principle, to exercise personal prejudice as other deciders.

The answer to the question who should decide does not yet exist. There may be no simple answer. Instead, we may need to create a complex of procedure and substance containing the following elements:

1. A statement of objective criteria. It must ask whether the intervention will cause or continue suffering, produce only pain, be of absolutely no use except to prolong the process of death, or have any possible chance of permitting a sapient or cognitive existence of any sort.

2. There must be a process of consultation. These sorts of decisions must, at least within the medical service, be open discussions between physicians, nurses, social workers, and parents. There must be the participation of some equals so that the medical hierarchy does not create the fiction of joint decision-making with the fact of singular power.
3. In all cases, there should be the availability of a forum to discuss prognosis, evaluate evidence, and attempt to weigh and balance the difficult options under consideration.
4. There must be the possibility of review in a timely fashion, appeal within the hospital itself, and access to the courts in special circumstances.

Despite these criteria, there must also be recourse to the courts in especially troubling and conflict-ridden situations. Whether support by a "right to privacy" or by the reasonable evaluation of commonly held norms of what humans expect and desire from existence, the protections for the life of the child must be maximal while the possibility of permitting death must be maintained.

In testimony in the *Quinlan* case, one doctor described her care as "judicious neglect."[21] He stated that terminally ill patients were often treated in such fashion when it "does not serve either the patient, the family, or society in a meaningful way to continue treatment."[22] That same language can apply to neonates. The same options must be available.

The key to many of the questions posed may be answered by a mandatory team approach to decision-making that focuses exclusively on the individual child and includes parents in the process of discussion. This will, of course, leave very many difficult problems unaddressed. What are the social policy consequences of keeping alive a child whose parents have rejected it? Can parents abandon a child kept alive against their wishes to the state? Will the state assume a greater share of the costs of caring for these children? Will the provision of decent prenatal care, which we know prevents many of these problems, now be seen as imperative? The principle that a child can be deprived of its only life is a narrow window in the wall of treatment.

Notes and References

[1]Oliver Wendell Holmes, *The Common Law* 35 (1881).

[2]*De Praerogativa Regis*, 17 Edw. 2, C.9 (1324).

[3]Revised Laws of Illinois, 1899, pp. 131–137.

[4]*See Griswold v. Connecticut*, 381 US 479 (1965).

[5]*See Roe v. Wade*, 410 US 113 (1973).

[6]*See In re Yetter*, 62 Pa. D & C.2d 619 (C.P. Northhampton County 1973); *In re Quinlan*, 355 A. 2d 647 (NJ 1976); *Superintendent of Belchertown State School v., Saikewicz*, 370 N. E. 2d 417 (MA 1977).

[7]*Quinlan, supra* note 6.

[8]*Saikewicz, supra* note 6.

[9]*Quinlan, supra* note 6, at 664.

[10]*Id.*, at 663.

[11]*Id.*, at 664.

[12]*Id.*, at 663.

[13]*Eisenstadt v. Baird*, 405 US 438, 453 (1972).

[14]*Saikewicz, supra* note 6, at 431.

[15]*In re Carson*, 241 NYS 2d 288, 289 (NY Sup. Ct. 1962).

[16]*Saikewicz, supra* note 6, at 431.

[17]*Id.*, at 422.

[18]*In re Storar*, 420 N.E. 2d 64 (NY 1981).

[19]*Bellotti v. Baird*, 443 US 622 (1979).

[20]*Ginsberg v. New York*, 390 US 629, 639 (1968).

[21]*Quinlan, supra* note 6, at 657.

[22]*Id.*

Section III

Images of the Abandoned

The Tyranny of the Normal

Leslie A. Fiedler

I embark upon this essay with a sense that I am an amateur writing for professionals, a dilettante addressing the committed. I am not a doctor or a nurse or a social worker, confronted in my daily rounds with the problem we discuss here; not even a lawyer, philosopher, or theologian trained to deal with its moral and legal implications. I am only a poet, novelist, critic—more at home in the world of words and metaphor than fact, which is to say, an expert, if at all, in reality once removed. I feel therefore like a Daniel in the Lions' Den, or perhaps I mean rather a Lion in the Daniels' Den— more victim than witness. Yet you have called upon me to testify and, with whatever fear and trembling, I have accepted.

Indeed, I have been asked several times recently to write on these topics because, I presume, not so very long ago I published a book called *Freaks: Myths and Images of the Secret Self*. I was primarily interested in exploring (as the subtitle declares) the fascination of "normals" with the sort of congenital malformations traditionally displayed at Fairs and Sideshows: and especially the way in which such freaks are simultaneously understood as symbols of the absolute Other and the essential Self. But in the long and difficult process of putting that study together (it took almost six years of my life and led me into dark places in my own psyche I was reluctant to enter), I stumbled on the topic treated here, "the care of imperiled newborns."

How dangerous a topic this was I did not realize, however, until just before my book was due to appear. I was at a party, in fact, celebrating the imminent blessed event when I

151

mentioned offhand to one of my fellow-celebrants, a young man who turned out to be an MD, one of the discoveries I had made in the course of my research. It seemed to me, I innocently observed, that in all probability more "abnormal" babies were being allowed to die (in effect, being killed) in modern hospitals than had been in the Bad Old Days when they were exposed and left to perish by their fathers. And I went on to declare that in my opinion at least this was not good; at which point, my interlocutor screamed at me (rather contradictorily, I thought) that what I asserted was simply not true—and that, in any case, it was perfectly all right to do so. Then he hurled his Martini glass at the wall behind my head, barely missing me, and stalked out of the room.

He did not stay long enough for me to explain that our disagreement was more than merely personal, more than the traditional mutual misunderstanding and distrust of the scientist and the humanist. Both of our attitudes, I wanted to tell him, had deep primordial roots: sources far below the level of our fully conscious attitudes and the facile rationalizations by which we customarily defend them. It is, therefore (I would have explained to him as I am now explaining here), to certain ancient myths and legends that we must turn—and here my literary expertise stands me in good stead—to understand the deep ambivalence toward fellow creatures who are perceived at any given moment as disturbingly deviant, outside currently acceptable physiological norms. That ambivalence has traditionally impelled us toward two quite different responses to the "monsters" we beget. On the one hand, we have throughout the course of history killed them—in the beginning ritually at least, as befits divinely sent omens of disaster, portents of doom. On the other hand, we have sometimes worshipped them as if they were themselves divine, though never without overtones of fear and repulsion. In either case, there was a sense of wonder or awe attached to anomalous humans: a feeling that such deviant products of the process by which we continue the species are mysterious, uncanny, and hence "taboo."

Though it may not be immediately evident, the most cursory analysis reveals that not merely have the primitive wonder and awe persisted into our "scientific," secular age;

but so also have the two most archaic ways of expressing it. We still, that is to say, continue to kill, or at least allow to die, monstrously malformed neonates. We euphemize the procedure, however, disguise the superstitious horror at its roots, by calling what we do "the removal of life-supports from nonviable *terata*" (*terata* being Greek for "monsters"). Moreover, thanks to advanced medical techniques, we can do better these days than merely fail to give malformed premies a fair chance to prove whether or not they are really "viable." We can detect and destroy them before birth, even abort them wholesale when the occasion arises, as in the infamous case of the Thalidomide babies of the 60s. That was a particularly unsavory episode (which we in this country were spared), since the phocomelic infants carried by mothers dosed with an antidote prescribed for morning sickness were in the full sense of the word "iatrogenic freaks." And the doctors who urged their wholesale abortion (aware that less than half of the babies lost would be deformed, but why take chances?) were in many cases the very ones who had prescribed the medication in the first place.

To be sure, those responsible for such pre-infanticide did not confess, were, indeed, not even aware—though any poet could have told them—that they were motivated by a vestigial primitive fear of the abnormal, exacerbated by guilt. They sought only, they assured themselves and the rest of the world, to spare years of suffering to the doomed children and their parents; as well as to alleviate the financial burden to those parents and the larger community, which would have to support them through what promised to be a nonproductive lifetime. We do, however, have at this point records of the subsequent lives of some of the Thalidomide babies whose parents insisted on sparing them (a wide-ranging study was made some years later in Canada), which turned out to have been, from their own point of view at least, neither notably nonproductive, nor especially miserable. Not one of them, at any rate, was willing to confess that he would have wished himself dead.

But such disconcerting facts do not faze apologists for such drastic procedures. Nor does the even more dismaying fact that the most whole-hearted, full-scale attempt at teratacide occurred in Hitler's Germany, with the collaboration, by the way, of not a few quite respectable doctors and teratologists, most notably a certain Etienne Wolff of the University of Strasbourg. Not only were dwarfs and other "useless people" sent to Nazi extermination camps, and by logical extension, parents adjudged (on grounds since totally discredited) likely to beget anomalous children sterilized. But other unfortunate human beings regarded—at that time and in that society—as undesirable deviations from the norm were also destroyed: Jews and Gypsies, first of all, with blacks, Slavs, and Mediterraneans presumably next in line. It is a development that should make us aware of just how dangerous enforced physiological normality is when the definition of its parameters falls into the hands of politicians and bureaucrats. And into what other hands can we reasonably expect it to fall in any society we know or can imagine in the foreseeable future?

Similarly, as the ritual slaughter of Freaks passed from the family to the State, the terrified worship of Freaks passed from the congregation to the audience, from the realm of Religion to that of entertainment and art. From the beginning, the adoration of Freaks was in the Western world a semirespectable, even an underground, Cult. Think, for instance, of the scene in Fellini's *Satyricon,* in which a Hermaphrodite is ritually displayed to a group of awestricken onlookers, who regard him not just as a curiosity though not quite as a god either. After all, the two reigning myth-systems of our culture, the Hellenic and the Judeo-Christian, both disavowed the portrayal of the divine in this guise; regarding as barbarous or pagan the presentation of theriomorphic, two-headed, or multilimbed divinities. By the Hebrews the portrayal of the divine was totally forbidden, while the Greeks portrayed their gods in idealized human form, i.e., the "Normal" raised to its highest power.

Yet there would seem to be a hunger in all of us, a need (without fulfilling which we would somehow feel ourselves less than human) to behold in wonder our mysteriously anomalous brothers and sisters. For a long time, this quasireligious need was satisfied in Courts for the privileged few, at fairs and sideshows for the general populace, by collecting and exhibiting Giants, Dwarfs, Intersexes, Joined Twins, Fat Ladies, and Living Skeletons. Consequently, even in a world that grew ever more secular and rational, we could still continue to be baffled, horrified, and moved by Freaks, as we were able to be by fewer and fewer other things once considered most sacred and terrifying. Finally though, the Sideshow began to die, even as the rulers of the world learned to be ashamed of their taste for human "curiosities"; but fortunately not before their images had been preserved in works of art, in which their implicit meanings are made manifest.

Walk through the picture galleries of any museum in the western world and you will find side by side with the portraits of Kings and Courtesans equally immortal depictions of the Freaks they kept once to amuse them (by painters as distinguished as Goya and Velasquez). Nor has the practice died out in more recent times, carried on by artists as different as Currier and Ives (who immortalized such stars of P. T. Barnum's sideshow as General Tom Thumb) and Pablo Picasso (who once spent more than a year painting over and over in ever shifting perspectives the dwarfs who first appeared in Valesquez' *Las Meninas*). Nor did other popular forms of representation abandon that intriguing subject. Not only have photographers captured on film the freaks of their time, but after a while they were portrayed in fiction as well. No sooner, in fact, had the novel been invented, that it too began to portray the monstrous and malformed as objects of pity and fear—and, of course, wonder, always wonder. Some authors of the nineteenth century, indeed, seem so freak-haunted that we can scarcely think of Victor Hugo, for instance, without recalling the grotesque Hunchback of Notre Dame, any more than we can recall Charles Dickens without thinking of his monstrous dwarf, Quilp, or our own Mark Twain without remembering "Those Incredible Twins." And the tradition is continued by modernists like John Barth and Donald Barthelme and Vladimir Nabokov, who, turning their backs on almost all the other trappings of the conventional novel, reflect still its obsession with Freaks.

In the twentieth century, however, the images of congenital malformations are, as we might expect, chiefly preserved in the artform invented during that century, the cinema: on the one hand, in Art Films intended for a select audience of connoisseurs, like the surreal fantasies of Fellini and Ingmar Bergmann; and on the other, in a series of popular movies from Todd Browning's thirties' masterpiece, *Freaks,* to the more recent *Elephant Man,* in its various versions for large screen and small. Browning's extraordinary film was by no means an immediate success, horrifying, in fact, its earliest audience and its first critics, who drove it from the screen and its director into early retirement. But it was revived in the sixties and has ever since continued to be replayed all over the world, particularly in colleges and universities. And this seems especially appropriate since in the course of filming his fable of the Freaks' revenge on their "normal" exploiters, Browning gathered together the largest collection of show-freaks ever assembled in a single place: an immortal Super-Sideshow, memorializing a popular artform now on the verge of disappearing along with certain congenital malformations it once "starred" (now routinely "repaired") like Siamese Twins.

Precisely for this reason, perhaps, we are these days particularly freak-obsessed, as is attested also by the extraordinary success of *The Elephant Man* on stage and TV and at the neighborhood movie theatre. The central fable of that parabolic tale, in which a Doctor, a Showman, the Press, and the Public contend for the soul of a freak, though it happened in Victorian times, seems especially apposite to our present ambivalent response to the fact of human abnormality. It serves indeed to remind us of what we now otherwise find it difficult to confess except in troubled sleep, though the arts have long tried to remind us of it: that those wretched caricatures of our idealized body image, which at first appear to represent what is most absolutely "Other" (thus reassuring us who come to gape that we are "normal") are really a revelation of what in our deepest psyches we suspect we recognize as the Secret Self. After all, not only do we know that each of us is a freak to someone else; but in the depths of our unconscious (where the insecurities of childhood and adolescence never die) we seem forever freaks to ourselves.

Perhaps it is especially important for us to realize that finally *there are no normals*, at a moment when we are striving desperately to eliminate freaks, to normalize the world. But this impulse represents a third, an utterly new response to the mystery of human anomalies—made possible only by the emergence of modern technology and sophisticated laboratory techniques. Oddly enough, however (and to me terrifyingly), it proved possible for experimental science to produce monsters long before it had learned to prevent or cure them. And to my mind, therefore, the whole therapeutic enterprise is haunted by the ghosts of those two-headed, three-legged, one-eyed chicks and piglets that the first scientific teratologists of the eighteenth century created and destroyed in their laboratories.

Nonetheless, I do not consider my enemies those first experimenters with genetic mutation, for all their deliberate profanation of a mystery dear to the hearts of the artists with whom I identify. Nor would I be presumptuous, heartless enough to argue—on esthetic or even moral grounds—that congenital malformations under no circumstances be "repaired," or if need be, denied birth, to spare suffering to themselves or others. I am, however, deeply ambivalent on this score for reasons, some of which I have already made clear, and others of which I will try to make clear in the conclusion of these remarks. I simply do not assume (indeed, the burden of evidence indicates the contrary) that being born a freak is *per se* an unendurable fate. As I learned reading scores of biographies of such creatures in the course of writing my book, the most grotesque among them have managed to live lives neither worse nor better than that of most humans. They have managed to support themselves at work that they enjoyed (including displaying themselves to the public); they have loved and been loved, married and begot children— sometimes in their own image, sometimes not.

More often than not, they have survived and coped; sometimes, indeed, with special pride and satisfaction because of their presumed "handicaps," which not a few of them have resisted attempts to "cure." Dwarfs, in especial, have joined together to fight for their "rights," one of which they consider to be *not* having their size brought up by chemotherapy and endocrine injections to a height we others call "normal," but they refer to, less honorifically, as "average." And I must say I sympathize with their stand, insofar as the war against "abnormality" implies a dangerous kind of politics, which beginning with a fear of difference, eventuates in a tyranny of the Normal. That tyranny, moreover, is sustained by creating in those outside the Norm shame and self-hatred-particularly if they happen to suffer from those "deformities" (the vast majority still) we cannot prevent or cure.

Reflecting on these matters, I cannot help remembering not merely the plight of the Jews and Blacks under Hitler, but the situation of the same ethnic groups—more pathetic-comic than tragic, but deplorable all the same—here in theoretically nontotalitarian America only a generation or two ago. At that point, many Blacks went scurrying off to their corner pharmacy in quest of skin bleaches and hairstraighteners; and Jewish women with proud semitic beaks (how it has all changed since Barbra Streisand) sought plastic surgeons for nose-jobs. To be sure, as the case of Streisand makes clear, we have begun to deliver ourselves from the tyranny of such ethnocentric Norms in the last decades of the twentieth century; so that looking Niggerish or Kike-ish no longer seems as freakish as it once did, and the children of "lesser breeds" no longer eat their hearts out because they do not look like Dick and Jane in their Primers.

But the Cult of Slimness, that aberration of Anglo-Saxon taste (no African or Slav or Mediterranean ever believed in his homeland that "no one can be too rich or too thin") still prevails. And joined with the Cult of Eternal Youth, it has driven a population growing ever older and fatter, to absurd excesses of jogging, dieting, and popping amphetamines—or removing with the aid of plastic surgery those stigmata of time and experience once considered worthy of reverence. Nor do things stop there; since the skills of the surgeon are now capable of recreating our bodies in whatever shape whim and fashion may decree as esthetically or sexually desirable: large breasts and buttocks, at one moment, meagre ones at another. But why *not*, after all? If in the not so distant future, the grosser physiological abnormalities that have for so long haunted us disappear forever—prevented, repaired, aborted, or permitted to die at birth, those of us allowed to survive by the official enforcers of the Norm, will be free to become even more homogeneously, monotonously beautiful; which is to say, supernormal, however that ideal may be defined. And who except some nostalgic poet, in love with difference for its own sake, would yearn for a world where ugliness is still possible? Is it not better to envision and work for one where all humans are at least *really* equal—physiologically as well as socially and politically?

But, alas (and this is what finally gives me pause), it is impossible for all of us to achieve this dubious democratic goal; certainly, not in the context of our society as it is now and promises to remain in the foreseeable future: a place in which supernormality is to be had not for the asking, but only for the buying (cosmetic surgery, after all, is not included in Medicare). What seems probable, therefore, as a score of science fiction novels have already prophesied, is that we are approaching with alarming rapidity a future in which the rich and privileged will have as one more, ultimate privilege, the hope of a surgically, chemically, hormonally induced and preserved normality—with the promise of immortality by organ transplant just over the horizon. And the poor (whom, we are assured on good authority, we have always with us) will be our sole remaining Freaks.

Comment on "The Tyranny of the Normal"

Ruth Macklin

Moral and Esthetic Values

It is hard to know precisely what conclusions to draw from Leslie Fiedler's provocative remarks. Would we be correct in inferring that the medical profession's efforts to minimize the incidence of physical deformities is yet another instance of our culture's deference to high standards of physical beauty and perfection? Should we conclude that one way to deemphasize or try to change those standards is to cease making surgical interventions in cases of correctible deformities? Or is Fiedler simply ruminating on a general cultural phenomenon and the role of modern medicine in addressing that phenomenon? The answer is not clear.

The issue as posed by Fiedler sets up a tug-of-war between an intellectual conclusion we may be inclined to draw and our emotional responses. Intellectually, we may concur with Fiedler's view that reigning cultural standards of acceptable appearance and normalcy are narrow, discriminatory, and intolerant. Emotionally, however, we are likely to recoil at the thought that our own child or children might deviate significantly from those existing standards. One way to approach this tension is by distinguishing between two sorts of values—moral and esthetic.

It is widely acknowledged that moral values stand higher than esthetic ones in the hierarchy of values. Standards of

161

physical beauty and appearance are by and large esthetic values, but at some point they shade into moral considerations. As a case in point, consider the difference between two types of plastic surgery: cosmetic and reconstructive. The former may be thought of as primarily esthetic in its objective, although it is possible that the "best interest" of a child may be served by alterations intended to make the child's appearance (usually facial) more attractive and thereby enhance the child's social adjustment. Reconstructive surgery, on the other hand, has a direct moral as well as medical objective. It is to render a grossly deformed individual (whether resulting from birth anomaly or accident) acceptably normal, in order to enable the person to live a life containing a modicum of social pleasures. Yet it is quite clear, in ordinary life circumstances, that an infant or child who differs from others markedly in appearance will fail to enjoy an equivalent quality of life, and will perhaps even fall below a minimally decent standard of social and interpersonal satisfactions.

We must distinguish, therefore, between two different issues. The first is that of tolerance toward our fellow humans who differ physically, mentally, and behaviorally from standards of normalcy; a principle of tolerance requires us to develop attitudes of acceptance and support for those individuals. The second issue is the appropriateness of seeking to make deformed infants and children more like their normal counterparts. It is not at all inconsistent to adopt and promote the principle of tolerance and at the same time choose to act in the best interest of children with significant physical handicaps, for their own sake, intervening at an early age to improve their appearance.

The anomalies under consideration here are not, with very few exceptions, life-threatening conditions. Medical interventions in these cases are therefore not the sorts that involve life-and-death decisions for infants. The presence of a physical abnormality of the kind we are discussing here could not justify failing to provide nutrients or other life supports for afflicted newborns, nor could they be used as an excuse for refusing consent for another medical procedure necessary to preserve the infant's life. Although they are not life-threatening conditions, these abnormalities are nevertheless likely to affect the individual's quality of life, so whether or not to correct them medically or surgically is surely a moral issue.

Morally Acceptable Approaches

In response to Leslie Fiedler's presentation, then, we need to ask what approaches to deformed infants are morally acceptable? We may, along with Fiedler, deplore the esthetic standards of beauty and normalcy that reign in our culture; but what is the appropriate response? Is it reasonable to suppose that those standards can be changed by individual actions, such as foregoing surgical or other methods of repairing deformities? Experience as well as common sense suggest the fruitlessness of such individual actions. But even if that approach could succeed, albeit minimally, in eroding cultural standards of physical attractiveness, is it fair to the particular child to sacrifice him or her to that end? Almost every available ethical principle yields a resounding "no" in answer to that question. Probably the ethical principle most relevant to this issue is a version of Kant's categorical imperative: "People should never be treated merely as a means, but should always be treated as ends in themselves." To use a child, or physically deformed children generally, as a means to the hoped-for end of changing society's standards of physical attractiveness would be a violation of the Kantian moral imperative. It would be all the more unethical because children are vulnerable beings, deserving of our care and protection, and those born with physical or mental handicaps suffer from the further disadvantages their disabilities create.

It is one thing to conclude that in general, the best interests of infants and children are served by performing surgery or other treatments to improve their appearance or correct physical deformities. It is quite another to decide what are appropriate responses to parental refusals of recommended procedures in such cases. Regardless of the parents' reasons for deciding not to make their child more normal physically, the likely prospect is resort to the courts or to other legal mechanisms. In effect, this approach sanctions government intrusion into family matters, a result that some commentators deplore.

Overriding Parental Decisions

One who has taken a strong stand in rejecting any role of the state in such matters is Joseph Goldstein, a Professor of Law at Yale Law School. In an essay entitled "Medical Care for the Child at Risk: On State Supervention of Parental Autonomy,"[1] Goldstein attacks the notion of state intrusion into parental decision-making, even in a range of cases where refusal of recommended medical treatment may result in the death of the child. His position in cases of non-life-and-death treatment is virtually absolute:

> When death is not a likely consequence of exercising a medical care choice there is no justification for governmental intrusion on family privacy; nor is there justification for overcoming the presumption of either parental autonomy or the autonomy of emancipated children. [2]

Goldstein's view, as stated here and elsewhere, rests on a number of initial premises: that the moral values involved in these cases are irreducibly subjective; that the concept of "the best interest of the child" is at best, flawed, and at worst, meaningless; and that no one is better qualified than parents to make decisions for their children.

One case Goldstein discusses at length is that of Kevin Sampson, a fifteen-year old boy who had a large, disfiguring growth on his face (the condition known as neurofibromatosis, more popularly, "Elephant Man's disease"). Kevin's mother refused to allow any necessary blood transfusions during surgery, and after testimony by doctors who recommended the operation, a judge declared Kevin a neglected child and ordered a series of operations. Goldstein argues that Kevin's mother was best qualified to determine the meaning of a "normal and happy existence" for her son; and that "the power of the state must not be employed to reenforce prejudice and discrimination against those who are cosmetically or otherwise different."[3] From these remarks, it appears that

Goldstein views the values at stake as mere matters of taste, of personal preference, and of lifestyle choices, rather than as straightforwardly moral values. Because of his denial that any objective meaning can be ascribed to the notion of the "best interest" of the child, Goldstein rejects any appropriate role for the state, and accuses the judge of arrogance, of performing an act of conjury, and of qualifying himself as "prophet, psychological expert, and all-knowing parent."[4] Yet Goldstein fails to report the finding of a psychologist who testified that Kevin was extremely dependent, and moreover, that his "deformity prompted social isolation and compelled his withdrawal from school, leaving him virtually illiterate in his teens."[5]

"Best Interest" of the Child

It does not require much additional evidence to conclude that, on any reasonable theory of "best interest" in our society, illiteracy is clearly against a person's best interest. To be virtually illiterate as a teenager, along with the burden of gross facial disfigurement, is an almost certain guarantee of personal misery. These factors should be weighed against the probable accuracy of Goldstein's judgment that

> . . . *under the protective cloak of family privacy, a loving, caring, accepting, autonomous parent had somehow been able to nurture in Kevin a "healthy personality." Kevin, after all, had developed in spite of state-reenforced prejudice and discrimination against the physically different in school, health agency, and court.*[6]

This is not the place to debate the wisdom of the court's ruling, given the complexity of the Sampson case, the age of the minor, the chances of recurrence of the growth and other pertinent details. My point is to explore Goldstein's assessment of the situation and the argument he mounts against state intervention that seeks to attain the best interest of the child. Goldstein's analysis affords us the opportunity to trace out some of the provocative notions put forth by Leslie Fiedler. If the case of Kevin Sampson, along with others Goldstein discusses, serves to illustrate the likely practical implications of Fiedler's views, I, for one, have very grave moral reservations about those views. Perhaps Fiedler himself is not prepared to go as far as Goldstein; he has not told us, so it may be unfair to attribute even that possible implication to him.

It is apparent that a major difficulty with an approach to physical deformities that mandates corrective surgery or other medical treatments is the prospect of involving courts or other legal devices to override parental refusals of treatment. Yet this may be what is required in order to provide for the best interests of children. This is not a happy solution to the problem posed by correctible deformities in infants and children. But it is in the nature of a moral dilemma that there are no easy answers.

Notes and References

[1]The essay originally appeared in *The Yale Law Journal*, **86** (March 1977). It is reprinted, with slight revisions, in *Who Speaks for the Child: The Problems of Proxy Consent*, Willard Gaylin and Ruth Macklin, eds., New York, Plenum Press, 1982, pp. 153–188.

[2]Goldstein, in Gaylin and Macklin, pp. 178–179.

[3]*Ibid.*, p. 183.

[4]*Ibid.*, pp. 181–182.

[5]Stuart J. Baskin, "State Intrusion into Family Affairs," *Stanford Law Review* **26** (1974): 1398, n. 74.

[6]Goldstein, p. 181.

Section IV

Caretakers: Images and Attitudes

Consensus and Controversy in the Treatment of Catastrophically Ill Newborns

Report of a Survey

Betty Wolder Levin

Introduction

Questions concerning the treatment of newborns have ceased being matters of private decision alone, and emerged as public issues. Advances in biomedical technology and practice now enable clinicians to prolong the lives of many infants, including some who will be left with catastrophic impairments. For these babies, the traditional bioethical norms that have guided clinical practice—"to preserve life" and "to do no harm"—seem to be in contradiction. New clinical norms have evolved that sanction the withholding of possible treatments from some of these critically ill newborns. The "Baby Doe Directives" were the first attempt of the Federal government to regulate such treatment practices. However, there have been few studies of the norms guiding neonatal decision-making and of their implications for the establishment of guidelines to regulate such decisions.

This paper reports the results of a survey about treatment decisions for catastrophically ill infants conducted soon after

169

the first active plans to enforce the directives were announced in the spring of 1983. The questionnaires were distributed to people concerned with the fate of catastrophically ill newborns, including many who care for such babies, at a conference and at a major medical center.

The purpose of the paper is to examine and report on the results of the survey and to discuss two sets of questions. The first concerns the investigation of norms to guide decision-making in neonatology. The recommendations of respondents about whether a choice to give or withhold specific treatments would be "best for the baby" in four hypothetical cases are reported. Information on consistency and variation between cases and treatments, and associations with the background characteristics of respondents are presented. The following questions are discussed: Is there agreement about the manner in which treatment decisions are to be made? Are there patients or treatments for which there is consensus? What are the areas of controversy?

The second set of questions concerns the establishment of guidelines for decision-making. Respondents were asked if they thought that specific treatments would be required by the Baby Doe Directives, and how they would rate these treatments on an "ordinary/extraordinary scale." This was followed by some general questions on who should make decisions, what types of guidelines should exist, and what factors should be considered important for decision-making. In this article, the following questions are examined: How were the directives interpreted? Did they provide a clear standard for decision-making? Is the "ordinary/extraordinary" distinction useful for framing guidelines? What other implications can be drawn about guidelines from the results of the survey?

Background

Advances in biomedical technology and practice have enabled clinicians to prolong the lives of critically ill patients. For most patients, treatment is provided and it is of clear benefit. For others, however, treatment may lead to a prolongation of the dying process or to a very poor quality of life. Therefore, it has been asserted that in response, clinical norms about treatment have changed. A core concept of "classical medical ethics"—an absolute norm to preserve human life—has changed to an "ethic of responsibility" in which other values may be weighed against the value of life in deciding medical treatment.[1] In some cases, decisions are made to withhold treatment.

Although much has been written by clinicians, philosophers, lawyers, and others concerning how the critically ill in general and impaired neonates in particular *should* be treated, there has been relatively little sociological study of the attitudes and norms that guide *actual* treatment decisions. Diana Crane conducted the most comprehensive survey of the treatment of the critically ill. She distributed questionnaires with hypothetical vignettes describing neonatal and older patients and asked respondents to indicate whether they would provide specific treatments for these hypothetical patients. She concluded that physicians are moving from a physiological to a social definition of life with a growing consensus to treat some patients and to withhold treatment from others.[2] Other surveys, such as those by Shaw,[3] McKilligin,[4] Todres,[5] and Jonsen,[6] articles by clinicians such as Duff and Campbell,[7] and a sociological study by Anspach[8] have also shown that clinicians choose to actively treat most neonates, while withholding treatment from others with lethal conditions or a poor prognosis.

Previous studies have primarily focused on characteristics of *patients* associated with decisions "to treat" or "not treat." However, they also show that clinicians differentiate between the characteristics of *treatments* in making treatment decisions. Whereas earlier studies have provided valuable insights about the new norms concerning the importance of various *patient* characteristics, they provide less information about the norms concerning which characteristics of *treatments* are considered important in decisions to give or withhold specific treatments.

The survey discussed in this paper was conducted as part of an anthropological study of neonatal decision-making. The purpose of the study is to investigate the factors that affect the choice of medical treatment for catastrophically ill infants. The study has involved participant observation in a neonatal intensive care unit, observations in and interviews with staff working in other units, and an analysis of the clinical, legal, philosophical, and popular literature on decision-making in neonatology.

A major focus of the study is an investigation of the way that clinicians differentiate between treatments in making decisions. Rather than a global choice between giving *all* possible treatments to preserve life, and withholding *all* treatments, the clinicians caring for newborns make decisions concerning specific treatments, among alternate courses of treatment available, for a specific baby. In some cases they choose not to give specific treatments. In other words they are not choosing "to treat" or "not to treat," but rather are choosing from a range of possibilities which treatments to give and which to withhold.[9] In some cases, such decisions to withhold treatment lead to death; in other cases, the patient lives despite the fact that treatments have been withheld. For example, in the case of "Baby Jane Doe," neurosurgery was withheld, but antibiotics were given for meningitis. Without the antibiotics, she probably would have died within the first few weeks of life; now she may live for many years despite the "nontreatment" (withholding of neurosurgery) decision.

On the basis of the observations, it was concluded that one way that clinicians conceptualize the difference between treatments is in terms of their *aggressiveness*. Treatments that have such attributes as a large physiological effect, that are experimental, invasive, or involve the use of high technology, and/or that are costly in terms of staff time or monetary costs, such as the use of a respirator or neurosurgery, are considered more aggressive than other treatments, such as tube feedings and antibiotics that do not have such attributes. In those cases in which clinicians withhold some treatments, they generally withhold the more aggressive, and provide the less aggressive treatments. This is frequently reflected in discussions of cases in the clinical literature and also in the data obtained from other surveys.[10]

The concept referred to here as "aggressiveness" is a ranking of treatments based on the characteristics of treatments themselves, independent of the presumed benefit of the treatment for a patient. This appears to be a widely shared concept in the clinical setting, which is, however, often confused with another concept that involves differentiating between treatments according to their presumed benefit to the patient. The same set of terms has been applied to treatments differentiated on the basis of the characteristics of treatments and those differentiated on the basis of benefit to the patient. At one extreme, treatments are referred to as "extraordinary," "heroic," or "aggressive," whereas at the other end of the continuum, they are referred to as "ordinary" or "conservative".

During the past decade, clinicians and members of the general public have become increasingly aware of the problems involved in making treatment decisions for catastrophically ill newborns. It has been suggested that guidelines be established for making such decisions.

Many discussions concerning standards for giving or withholding treatment have been based on a distinction between "ordinary" and "extraordinary" care.[11] The terms, developed in Catholic moral theology, have been used in many contexts, including some Court decisions concerning the care of the critically ill.[12] The terms are defined in the literature as dichotomous, and according to most philosophers their use should be based only on the expected benefit of the treatment for a particular patient. It is usually assumed that individuals would agree about which treatments are ordinary and which are extraordinary for a particular patient.

Observation of the use of the terms in the clinical setting suggested that treatments are seen to vary along a continuum from ordinary to extraordinary. Often the terms are used to refer to distinctions based on characteristics of treatments, here defined as "aggressiveness" rather than distinctions based primarily on characteristics of patients. Also, there is not always agreement about which treatments are ordinary and which are extraordinary. In fact, in some cases some clinicians refer to treatments as ordinary, whereas others refer to the same treatments as extraordinary for the same patient.

The Baby Doe Directives were issued by the federal government in an attempt to set standards for treating handicapped newborns. The directives issued in March of 1983 stated that:

> Under section 504 it is unlawful for a recipient of federal financial assistance to withhold from a handicapped infant nutritional sustenance or medical or surgical treatment required to correct a life-threatening condition if: (1) the withholding is based on the fact that the infant is handicapped; (2) the handicap does not render the treatment or nutritional sustenance medically contraindicated.[13]

The directives required that signs about the policy be posted in all neonatal units giving an "800" number that could be called to report alleged violations and initiate investigations by federal representatives.

Many clinicians and others who were familiar with the issues involved in decision-making, objected to the directives. They claimed that the regulations had no legal basis, that they were unclear; and that they would probably require that many treatments be given to babies that would not be in their best interests.

Methods

Questionnaires were distributed to the 251 people who attended the conference entitled "Which Babies Shall Live: Humanistic Dimensions of the Care of Imperiled Newborns" presented by The Hastings Center and Montefiore Medical Center, held in New York City on April 6, 1983.[14] In late April and May 1983, a second version of the questionnaire was distributed at the Columbia-Presbyterian Medical Center to the 129 physicians, nurses, and other professional staff who work in the neonatal intensive care unit, and to the 52 participants attending a Department of Obstetrics and Gynecology Retreat. Although this is not a sample of a defined population, it does provide a set of responses from a number of people, most of whom are very knowledgeable about treatment decisions for catastrophically ill newborns. The first portion of the questionnaire, based on the work of Diana Crane, presented hypothetical vignettes of cases of newborns with four critical conditions—Down's syndrome with duodenal atresia, anencephaly, trisomy-13, and extreme prematurity (25 weeks gestational age). For each case, respondents were presented with a list of treatments, each of which they were to assume would increase the baby's chance of survival if given and they were asked to assume that the parents' views were the same as their own. Next respondents were asked to indicate: (1) if they thought it would be best to give or to withhold a number of treatment options for each baby by circling whether they would either definitely, probably, probably not, or definitely not recommend each treatment, (2) whether they thought the treatments would be required by the Baby Doe Directives, and (3) how they would rate each of the treatments on a scale from one (ordinary) to five (extraordinary) (*see* Appendix). Another portion of the questionnaire consisted of multiple choice questions concerning how decisions should be made for individual newborns, how policies should be set, and which factors should be important in making such decisions. In addition, there were questions requesting background demographic and other information on respondents. The questionnaires maintained the respondent's anonymity.

Characteristics of the Respondents

In total, 249 of the 432 questionnaires distributed were returned: 130 from the Conference, 97 from the Neonatal Intensive Care Unit (NICU), and 22 from the Obstetrics and Gynecology Department Retreat. The return rates for the response groups were 52, 75, and 42%, respectively.[15] The respondents from the neonatal intensive care unit and the Obstetrics and Gynecology Department retreat will together be referred to as the "Columbia respondents." Unless otherwise noted, all responses are combined and reported together. The responses of the conference and Columbia respondents were similar when responses of members of each occupational group were compared.

Of the respondents to the questionnaire, 30% were neonatal nurses, 18% were other nurses, 13% were neonatol-

ogists, 14% were other physicians, 18% worked in other
occupations or settings related to the delivery of health care
(medical social work, hospital clergy, hospital administration,
and so on) and 6% worked in law or journalism (most with a
special interest in health) and special education. Three percent worked at other occupations or their occupations were
unknown. Seventy-six percent of the respondents had professional experience working with catastrophically ill newborns.

Respondents ranged in age from 21 to 73 years, with an
average age of 37. Seventy-four percent of the respondents
were female, 26% male. Only the Columbia respondents were
asked about religion and religiosity. Of these, 28% identified
themselves as Protestant, 41% as Catholic, 17% as Jewish, 7%
as other, and 7% as none. Eleven percent characterized themselves as deeply religious, 60% as moderately religious, 26%
as indifferent to religion, and 3% as opposed to religion. The
response rate per item was generally high. Reported per-
centages are based on answers from at least 240 responses,
unless otherwise noted.

Results

First, the results are provided for each hypothetical case.
(Please see the Appendix for the full vignettes and treatment
choices for each case.) The discussion of each case includes
comments on consensus and controversy. Following Diana
Crane,[16] "consensus" is defined as agreement of at least 75%
of the respondents.[17] There is said to be controversy or lack of
consensus when less than 75% of the respondents were in
agreement. Second, there is a comparison of the treatments
recommended between cases. This includes a discussion of
the association of differences in treatment recommendations
with differences in respondent characteristics including
occupation, religion, and age. This is followed by findings on
consistency and variation concerning the interpretation of the
Baby Doe Directives, and about categorizations of treatments
as ordinary and extraordinary. Finally, responses to some
general questions on decision-making are presented.

Case by Case Summary of Results[a]

Baby "A": A Baby with Down's Syndrome and Duodenal Atresia

The Case of Baby "A", patterned after the case of "Baby Doe" of Bloomington, Indiana, represents the type of infant who is likely to have a moderate degree of mental retardation and also has a life-threatening defect that would be routinely repaired for a baby who did not have other impairments. There was consensus that such a baby should receive feedings and relatively routine surgery to repair such a defect. Fewer people would recommend other treatments, such as kidney dialysis, felt to be more extraordinary (*see* Table 1).

In general, most respondents felt that more extraordinary treatments would be required for Baby "A" under the "Baby Doe Directives" than they themselves would recommend. Since the case of Baby "A" was patterned after the case that inspired the Federal Directives, it seemed surprising to many that 10% of the respondents did not think that intestinal surgery would be required for Baby "A." In addition to those who did not think that it was required, 24% were uncertain whether surgery would be required. This group included neonatologists and neonatal nurses.

Baby "B": A Baby with Anencephaly[18]

The Case of Baby "B," a baby born without an upper brain, represents a baby with severe retardation and a terminal condition.[19] Most anencephalic babies die within the first few days of life; practically all die within the first few weeks. There was consensus to provide feeding for such a baby if the baby could suck, but to give not "heroic" treatments that would have prolonged the dying process. There was less agreement about the advisability of such "nonheroic" treatments as tube feedings and antibiotics that could be seen either as providing comfort or as prolonging the dying process. This case showed the largest disparities between what respondents thought were best and what they thought were required by the federal directives (*see* Table 1).[20]

[a]See appendix for description of each case.

Baby "C": A Baby with Trisomy-13[21]

Respondents made a clear distinction between the highly uncertain situation of a baby with multiple anomalies that could possibly be correctable, represented by Baby "C" before diagnosis, and the situation of a baby with a clear diagnosis of a lethal defect, as represented by the case of Baby "C" after diagnosis. Even though the baby's anomalies were suggestive of trisomy-13, most respondents would have recommended all of the treatments listed before chromosomal analysis. After a definitive diagnosis of this lethal anomaly, there was consensus that most treatments should be withheld, even though the baby's life might be prolonged for a period of time. However, most respondents felt that such treatments would be required by the Federal Directives even after a definitive diagnosis. Respondents categorized a number of treatments as ordinary before the chromosomal analysis, while categorizing those same treatments as extraordinary after the analysis (*see* Table 1).

Table 1   Treatments Respondents Would Recommend, Treatments Respondents Thought Were Required, and Ratings on the Ordinary/Extraordinary Scale
$n = 249^a$

Condition and treatment	Would recommend, %	Thought required, %	Mean score, ordinary/ extraordinary
Baby "A"—Down's syndrome and duodenal atresia			
Intravenous feedings	91	98	1.3
Antibiotics	88	95	1.5
Surgery for intestinal defect	87	90	2.2
Cardiac catheterization	71	76	2.9
Open heart surgery	59	69	3.7
Kidney dialysis	28	57	4.3
Baby "B"—anencephalic			
Feedings by mouth	76	90	1.9
Tube feedings	56	82	2.6

(*continued on next page*)

Table 1   *(continued)*

Condition and treatment	Would recommend, %	Thought required, %	Mean score, ordinary/ extraordinary
Antibiotics	32	75	3.1
Resuscitation in the delivery room	13	63	4.0
Cardiac catheterization	3	31	4.6
Arrest page	2	47	4.7
Open heart surgery	2	26	4.8
Baby "C"—Multiple anomalies (before chromosomal analysis)			
Nutrition and fluids	90	97	1.6
Antibiotics	81	96	1.9
Resuscitation	76	95	2.0
Respirator	65	91	2.4
After chromosomal analysis (trisomy-13)			
Nutrition and fluids	85	93	1.9
Antibiotics	60	83	2.8
Respirator	24	68	4.1
Surgery for cleft palate	14	57	4.0
Cardiac catheterization	13	49	4.5
Arrest page	10	51	4.6
Open heart surgery	8	46	4.7
Baby "D"—Small premature baby with IVH			
Nutrition and fluids	93	98	1.6
Suctioning	92	96	1.7
Increased respirator settings	53	78	2.3
Resuscitation in DR	64	88	2.3
Pressors	39	66	3.5
Arrest page	26	55	4.1
Kidney dialysis	13	42	4.6

[a]For Baby "C" and for questions on resuscitation in the delivery room and an arrest page for Baby "B," $n = 119$; actual base varies slightly depending on the number of ineligible answers (in all cases, ineligible answers less than 5% of total n).

Baby "D": A Very Premature Baby

This case represents the type of baby who is likely to have serious lasting impairments, but who might have only mild impairments or be completely normal. There was consensus to give treatments, such as nutrition, fluids, and suctioning categorized as ordinary, even after bleeding in the brain was diagnosed. There was controversy concerning resuscitation of such a baby at birth and about increasing the respiratory settings for such a baby after bleeding in the brain was detected, although there was consensus that such treatments would be required. There was consensus that kidney dialysis, a treatment categorized as extraordinary, should *not* be provided although there was a lack of consensus concerning whether kidney dialysis would be required by the directives (*see* Table 1).

Premature babies constitute the vast majority of babies admitted to neonatal intensive care units. A very premature, very low birth weight baby, like Baby "D", would have been considered nonviable as recently as five years ago. Now some of these babies may be saved with intensive care. The prognosis for such babies who are diagnosed as having significant bleeding in the brain is generally poor, but since some do well, the outcome for any particular baby is uncertain. Decisions for such cases are among the most common "ethical decision-making" problems encountered in neonatal intensive care units. Many clinicians report that they find them among the most difficult because of the high degree of uncertainty involved.

Treatment Recommendations:
Similarity and Variation
Between Cases

In comparing the responses across cases, there are clear differences in the levels of treatment that respondents would recommend. The number of respondents who felt it would be best to recommend any particular treatment varied from case to case. This is illustrated by the fact that many more respondents would recommend each particular treatment for the baby with Down's Syndrome and duodenal atresia than would recommend those same treatments for the baby with trisomy-13 and a cleft palate (*see* Table 1).

Despite the variation in the percentage of respondents who would recommend treatments between cases, however, there was consistency in the ranking of treatments from case to case. For example, more respondents always recommended giving antibiotics than recommended doing cardiac surgery for every case for which both were options. This is illustrated by the fact that 89% of the respondents recommended antibiotics and 59% recommended open heart surgery for the baby with Down's syndrome, whereas only 60% recommended antibiotics and 8% recommended open heart surgery for the baby with trisomy-13. Those treatments most likely to be given were the least aggressive treatments,[22] whereas the most aggressive treatments were most likely to be withheld; those treatments that were intermediate in their level of aggressiveness were also intermediate in terms of the frequency with which respondents recommended that they be withheld.

The Pattern of Treatment Recommendations

The ranking of treatment choices according to the aggressiveness of treatment was reflected not only in variation in the treatment choices between cases, but was also reflected in the pattern of individual respondents' recommendations for treatment. If a particular respondent recommended withholding a particular treatment for a particular case, that respondent was likely to withhold all other, more aggressive treatments for that case as well, and, conversely, if a respondent recommended giving a particular treatment, that respondent was likely to recommend giving all other less aggressive treatments in that case.

The extent to which the pattern of treatment recommendations of individual respondents reflected the same ordering of treatments as the respondents as a group was measured by use of Guttman scale analysis. The coefficient of reproducibility for the treatment recommendations for each case exceeded 0.92, which indicates that for each of the cases, the respondents answers were highly correlated with the ordering of treatments in terms of aggressiveness. For those cases that included questions concerning resuscitation in the delivery room or treatment before a definitive diagnosis was established (treatments that would be given at a different, earlier time), the coefficient of reproducibility was higher when such treatments were excluded from the analysis. When only the treatments that would be given in the intensive care unit, after diagnosis, were analyzed, the coefficient of reproducibility was even higher. In all cases it was above 0.97, indicating a pattern of responses with a very strong degree of ordering according to aggressiveness.

A second measure, a "transposition" score, was also used to measure the deviation of individuals' choices of treatments from the sets of choices that would be predicted according to the hypothesized norm that all treatments withheld would be more aggressive than all treatments chosen.[23] In 87% of the cases, the respondents' pattern of treatment choices corresponded to the hypothesized clinical norm that less aggressive treatments would be given and more aggressive treatments would be withheld. The difference between an individual respondent's answers and the pattern of the respondents as a whole yielded a transposition score greater than 1 in less than 5% of the cases. The pattern of the treatment recommendations did not deviate from the expected pattern for any of the four cases for 68% of the respondents.

Background Differences and Variation in the Aggressiveness of Treatment

Each respondent was assigned "case aggressiveness scores" based on the highest ranked treatment recommended for each case[24] and an overall "aggressiveness score" calculated as the sum of the case aggressiveness scores for babies "A," "B," and "C." This yielded overall aggressiveness scores that ranged from 0 to 20. The mean overall aggressiveness score for all respondents was 10.2, the standard deviation was

3.7. The distribution of aggressiveness scores was approximately normal.

There were some significant[25] differences between professional and religious groups in the aggressiveness of treatments recommended, although there was also much individual variation. For most cases, Catholic respondents in general, and Catholic members of the same occupational group in particular, tended to have higher aggressiveness scores than Protestant or Jewish respondents. The average overall aggressiveness score of Catholics (10.7) was significantly higher than the overall aggressiveness scores of Protestants (9.0), whereas the average score of Jewish respondents (9.6) was intermediate. Catholics were significantly more likely to recommend aggressive treatments for both the baby with Down's syndrome and the anencephalic baby than Protestant or Jewish respondents. For example, although respondents of all religious groups were likely to recommend intestinal surgery for the baby with Down's syndrome, only Catholics were likely to recommend cardiac surgery. Protestants recommended significantly fewer treatments for the baby with trisomy-13, though for that case the average level of treatment recommended by Jewish and Catholic respondents was the same. The differences in the aggressiveness scores for the very small premature baby were not significant; for that case, Jewish respondents were more likely to recommend more aggressive treatment than either Catholics or Protestants.

The average overall aggressiveness score of neonatal nurses (10.8) was significantly higher than that of neonatologists (9.4). Neonatal nurses were significantly more aggressive than neonatologists in their treatment recommendations for the babies with anencephaly and trisomy-13; they were more likely to recommend giving such relatively less heroic treatments as tube feedings (76 vs 47%) and antibiotics (34 vs 12%) for the anencephalic baby and antibiotics (81 vs 18%) and increased respiratory settings (18 vs 5%) for the baby with trisomy-13. These can be seen as treatments intended to "provide comfort" rather than "cure." Like the doctors, they were unlikely to recommend the most aggressive treatments for these babies. For example, no neonatologist or neonatal nurse recommended heart surgery for an anencephalic baby. The difference between the treatment recommendations of NICU nurses and neonatologists for the baby with Down's syndrome and the very small premature baby were not significant. The pattern of recommendations of members of each professional and religious group corresponded to the overall pattern of giving less aggressive treatments while withholding more aggressive ones.

Although there were significant differences in aggressiveness scores when occupational and religious categories were compared at the group level, occupation and religion were not significantly associated with aggressiveness after adjustment by multiple regression for age, sex, professional experience, religiosity, and occupation. The entire set of variables explained only 15% of the total variation in aggressiveness scores. Only age was a statistically significant explanatory factor ($p < 0.05$), accounting for approximately 10% of the variation; older respondents were *less* aggressive in their treatment recommendations than younger respondents. The inclusion of the additional variables, including religion and occupation, did not significantly improve the model.[26]

This finding seems to contradict the widely held assumption that differences in professional and religious background account for the perceived differences in opinions regarding treatment. Here, although some of the mean aggressiveness scores between groups of respondents with different professional and religious affiliations were significantly different, these differences were not associated with differences at the individual level. Among respondents to this questionnaire, as commonly occurs among the members of the staffs of many neonatal intensive care units, being a nurse, being Catholic, and being young were all associated with each other, and all tended to be associated with more aggressive treatment recommendations. But, individual doctors and nurses, and Catholics and non-Catholics who were the same age did not differ significantly in their treatment recommendations. Indeed, many of the older, more experienced nurses were *less aggressive* in their recommendations than many of the younger resident physicians and neonatal fellows. Therefore, it does not seem to be differences in professional socialization or religious training *per se* that account for differences in the degree of aggressiveness.

Interpretation of the Baby Doe Directives: Consistency and Variation

For every treatment choice, respondents were more likely to think it would be required by the directives than they were to feel it would be best for the baby.[27] This disparity ranged from 3 to 45% (*see* Table 1). Twenty-one percent of the respondents felt that every treatment would be required in every case. However, most respondents thought that even under the directives, some treatments would not be required. In general, there was consensus that the treatments that were rated most ordinary would be required, but there was controversy about the treatments that were rated most extraordinary.

For *no* treatment was there consensus (more than 75% agreement) that it would not be required by the directives. However, in all cases except that of the Down's syndrome baby, more than half of the respondents felt that some treatments could be withheld. Opinions about which treatments would not be required varied from case to case. Almost three quarters of the respondents did not think that cardiac surgery would be required for an anencephalic baby, and only 11% thought that it was *definitely* required.

Many respondents indicated that they were confused about the proper interpretation of the directives by circling that many treatments were "probably" as opposed to "definitely" required or not required. Not only were they unsure whether the Baby Doe Directives would require treatments that they *would not* recommend, but, in addition, some respondents were not sure about whether they would be required to give treatments that they *would* personally recommend. For example, 52% of the respondents were not sure that cardiac surgery would be required for a baby with

Down's syndrome.[28] This included 24 respondents who thought that surgery would *not* be required even though they would recommend it as best for the baby.

In general, respondents were likely to think that all treatments that they themselves would recommend would also be required by the directives. In addition, most felt that some of the more aggressive treatments, which they would not personally recommend, would also be required. The pattern of responses reflected the same ranking of treatments according to aggressiveness as discussed above. The Guttman scores for the treatments thought required in each case were all above 0.95.

Ratings on the Ordinary/Extraordinary
Scale: Consistency
and Variation

There was much variation in the ratings assigned to particular treatments on the ordinary/extraordinary scale. Like the responses to the questions about which treatments respondents would recommend as best and which they thought were required, however, there was a high degree of consistency in the rank ordering of treatments; the coefficient of reproducibility in all cases exceeded 0.92. The ranking of treatments was also consistent from case to case. For example, respondents would rate surgery as more extraordinary than antibiotics within each case. They would rate each treatment as more extraordinary for the baby with trisomy-13 than they would rate the same treatments for the baby with Down's syndrome. When comparing between cases, however, a treatment might be rated as more extraordinary in one case than a less aggressive treatment was rated in another case. For example, many respondents rated antibiotics for the baby with trisomy-13 as more extraordinary than surgery for the baby with Down's syndrome (*see* Table 2).

The variation in ratings was great. Ratings ranged from 1 (the most ordinary) to 5 (the most extraordinary) for *every* treatment choice from the one judged, on average, to be most

Table 2 Comparison of Ratings on the Ordinary/Extraordinary
Scale by Patient Condition and Treatment[a]

Treatments and patient conditions	Ratings				
	Ordinary		Extraordinary		
	1	2	3	4	5
IV feedings/nutrition and fluids					
Down's syndrome and duodenal atresia	83	8	5	2	2
Trisomy-13 with cleft palate	63	16	4	7	10
Antibiotics					
Down's syndrome and duodenal atresia	71	17	7	2	3
Trisomy-13 with cleft palate	28	18	18	17	19
Surgery					
For duodenal atresia (Down's)	33	33	22	7	5
For cleft palate (trisomy-13)	4	9	18	22	47
Open Heart Surgery					
For baby with Down's syndrome	4	9	27	27	33
For baby with trisomy-13	2	0	6	11	81

[a]Numbers represent the percent of respondents who would assign each treatment a particular rating on the "ordinary/extraordinary scale." For treatments for the baby with Down's syndrome, $n = 249$. For treatments for the baby with trisomy-13, $n = 119$).

ordinary (feedings for the baby with Down's syndrome) to the one judged most extraordinary (open heart surgery for an anencephalic baby).

Although there was not perfect agreement, in general, there was a high correlation of the responses concerning which treatments each respondent would recommend as best for the baby and the ratings on the ordinary/extraordinary scale (*see* Table 3). In general, respondents recommended giving treatments they would rate as ordinary, while withholding those they would rate as extraordinary. Some people, however, recommended withholding treatments they rated as ordinary, while others recommended giving some of the treatments they rated as extraordinary.

The treatments ranked as the more aggressive were rated higher on the ordinary/extraordinary scale by individual respondents. This included both those respondents who would recommend each treatment and those who would not recommend each treatment. The difference between the philosophers "ideal" definition based on beliefs concerning the benefit of a particular treatment for a particular baby and the use of the terms by the respondents is illustrated by the ratings listed in Table three.

Table 3   Rating on the Ordinary/Extraordinary Scale by Recommendations Concerning Treatments for the Baby with Down's Syndrome[a]

Treatment and treatment recommendations		Ratings				
		Ordinary		Extraordinary		
		1	2	3	4	5
Intestinal surgery						
Would recommend	$n = 216$	37	36	19	6	2
Would not recommend	$n = 31$	7	10	45	16	23
Cardiac surgery						
Would recommend	$n = 142$	7	15	39	20	19
Would not recommend	$n = 97$	1	1	9	35	54
Kidney dialysis						
Would recommend	$n = 67$	10	15	21	22	31
Would not recommend	$n = 173$	1	3	6	16	74

[a]Numbers represent the percent of respondents within each treatment category (recommending or not recommending treatment) who would assign the treatment a particular rating on the "ordinary/extraordinary scale."

General Questions About Decision-Making

When asked *who* should make decisions, respondents were about equally split between thinking that decisions about individual newborns should be made by the baby's parents with the advice of professionals caring for the baby (47%) or by a joint decision of parents and professionals (48%). Very few felt that decisions should be solely medical decisions or reflect uniform policies (4%).

In response to a question on *how* policies should be set, almost half of the respondents felt that no policies should be set because all decisions should be made on a case by case basis (49%). A substantial number wanted broad guidelines by hospital ethics committees (29%). Few (8%) wanted broad guidelines from the Federal Government. No one responded that they wanted specific federal guidelines.

Respondents were asked to express their opinions about the *basis* for decisions by indicating which of 21 factors they felt should be important when deciding about the care of individual newborns. Most respondents circled severity of intellectual impairment, severity of physical impairment, amount of prolonged pain and suffering, and parents' wishes. Next, the respondents were asked to indicate the single factor that they believed should be the most important factor for decision-making. The same four factors were again singled out most frequently (*see* Table 4).[29]

There were significant differences between neonatal nurses and neonatologists concerning which factor should be the most important in making treatment decisions. More neonatologists cited the amount of intellectual impairment (45 vs 25%), whereas more neonatal nurses thought that the amount of prolonged pain and suffering should be the most important factor (28 vs 3%).[30]

It is interesting to note that there were larger differences between what the members of different professional groups said should be the most important factor in guiding treatment decisions than there were in almost all of the actual decisions about treatment recommendations. This may reflect a greater

Table 4   Importance of Specific Factors for Decision-making

Factor	Percent who thought it should be *the most* important factor	Percent who thought it should be *an* important factor
Severity of intellectual impairment	29	87
Amount of prolonged pain and suffering	20	77
Severity of physical impairment	15	72
Parents' wishes	15	69
Capacity to give and receive love	4	30
If chance of successful treatment is small	3	47
Uncertainty about extent of impairment	3	42
If nontreatment would be active euthanasia	3	21
Impact on parents	2	49
If treatments are heroic	2	33
Danger of lessening the "value of life"	2	25
If treatments are already started	2	18
Long term cost of caring for disabled child and adult	0	38
Financial burden to family	0	33
Impact on siblings	0	30
Availability of resources for other sick children	0	26
Ability of the parents to have other healthy children	0	20
Cost of neonatal intensive care	0	19
Availability of resources for other medical care	0	19
Availability of resources for other, non-medical social needs	0	18
Feelings of staff caring for baby	0	15
Other factors	1	6

influence of personal background or professional ideology on the respondents' conscious beliefs and justifications for behavior. The actual behavioral choices may be more strongly influenced by underlying clinical norms shared by nurses and physicians, shaped by familiarity with and acceptance of the decisions that are actually made in neonatal intensive care units.

Discussion

The responses to this survey reflect the fact that among people who are knowledgeable about treatment choices for catastrophically ill newborns, there is consensus about some aspects of decision-making, but controversy about others. In this section, areas of consensus and controversy will be examined and implications for the establishment of guidelines will be discussed.

What Are the Issues for Which There Are Norms That Lead to Consensus Regarding the Giving or Withholding of Treatment? What Are the Areas of Controversy?

The overall pattern of responses to the questionnaire reflects the fact that to many people who are familiar with the issue, treatment choice is not a simple decision of giving either "treatment" or "no treatment." Rather, decision-making reflects clinical norms that sanction giving some treatments while possibly withholding others from the range of possibilities in some cases. Evaluations are made concerning both characteristics of patients and of treatments in making such decisions.

Consideration of withholding treatment starts with an evaluation of the patient. If there is no question of lasting physical or intellectual impairment, aggressive treatment is always given. If it is thought that a baby might suffer from permanent physical and/or intellectual impairments that would compromise the baby's future quality of life even with maximal treatment, decision-makers consider both the severity of impairment and probability of impairment. Both the severity and probability of impairment are seen as continua. Treatments may also be seen as varying in their level of aggressiveness, and this in turn is associated with such attributes of treatments as their invasiveness, their power to affect physiology, whether they are experimental or considered part of standard practice, their availability, and their "costs" in terms of money and staff time, and in the amount of pain and suffering involved for the baby and the baby's family.

When making treatment decisions, decision-makers seek to find what they consider an appropriate level in terms of the aggressiveness of treatment for each patient. There is consensus among knowledgeable people that the level of aggressiveness of the treatment should depend on patient characteristics, such as the future quality of life. There is a high degree of consensus about the appropriateness of giving or withholding some of the least and some of the most aggressive treatment options for some types of cases. However, for others, there is less consensus.

At one end of the continuum of patients, one finds cases of babies for whom there will be no serious lasting impairments likely to impair future quality of life. Although no such cases were represented by hypothetical cases on this questionnaire, such cases comprise the majority of babies admitted to neonatal intensive care units. For such babies, there is general agreement that even very aggressive treatments should be provided.

At the other end of the continuum is the category of patients on this questionnaire with lethal conditions, such as the anencephalic baby and the baby with trisomy-13. Even for such babies, the choice is not between total "treatment" and "nontreatment," but instead involves decisions about providing some treatments while withholding others. There is consensus to provide such non-aggressive treatments as nutrition and fluids, at least when the baby can be fed by mouth. As the aggressiveness of treatments increase, the percentage of respondents who would give each treatment decreases, until there is again consensus not to give the most aggressive treatments, such as open heart surgery or an arrest page for patients who have no hope of prolonged survival. The choice of treatment affects the quality of life for these babies while they survive and also affects the timing of their deaths.

For those babies who fall in between the extreme of the continuum, for whom there is a high probability of impairment, but the expected impairment is not seen as extremely severe (for example, a baby with Down's syndrome and duodenal atresia) or alternately, there is a high degree of uncertainty, even though there may be very severe lasting impairments (for example, the baby with multiple anomalies or a very small premature baby who has had bleeding in the brain), the choice will be to give some treatments while withholding others. The level of aggressiveness of the treatments recommended will be higher for most respondents for such babies than for those with lethal conditions. But many respondents would not be as aggressive in their treatment of these babies as they would for a baby who had a high probability of having no serious lasting defects.

For some babies who have a high probability of a moderate impairment or for whom there are likely to be severe impairments but there is a significant amount of uncertainty, there is controversy about the appropriate level of treatment to give. There is also controversy about the use of some of the most aggressive treatments (such as organ transplants and extra corporeal membrane oxygenation, which are *very* invasive, costly, and/or experimental) for any baby, and also about the use of some relatively ordinary treatments (such as iv feedings) for some infants who are terminally ill.

In sum, there is consensus to consider both patient characteristics and treatment characteristics in making treatment choices. There is consensus about treatment decisions at the extremes, such as providing nutrition and fluids to all newborns who can suck (when feedings would not be physiologically harmful) and about withholding very aggressive treat-

ments, when they would only serve to prolong the dying
process. However, there is no consensus about many of the
treatment options that fall in between, and there is controversy among knowledgeable persons about some other
treatment options.

What Are the Implications of These Findings For Guidelines For Decision Making?

First, how were the Baby Doe Directives interpreted by
respondents to the questionnaire? In general, respondents
seemed to interpret the directives as meaning that more
aggressive treatments would be required for each baby than
they thought might be best, but that the directives would not
change the manner in which treatment decisions were to be
made. That is, responses indicate that the respondents felt
that even under the directives, decisions should reflect the
clinical norm that sanctions an evaluation of the patient and
treatment characteristics. Some respondents did not think
that they would be required to give some of the treatments
they would recommend. As indicated by the fact that many
respondents circled that treatments were "probably" rather
than "definitely" required or not required, many respondents
were not certain about exactly which treatments would be
required under the directives.

The interpretation of the directives by some respondents
was different from that of some of the advocates of the directives who felt that all possible treatments would be required and some of the critics of the directives who also felt
that the directives would not allow a decision to withhold
even futile treatments. To those people who thought all treatments were clearly required, it seemed surprising that a number of respondents did not think that they would be required
to give even treatments that the respondents themselves
would recommend.

New versions of the directives that were issued after this
survey was conducted, specified that "heroic treatments" that
would be "futile" because they would only serve to prolong

the dying process could be withheld, but that "supportive care" should always be provided, even to babies with terminal conditions. The new directives also stated that physicians should have discretion in making "medical decisions." At the same time, it was asserted that not all decisions made by physicians are medical decisions and that decisions based on the future quality of life are prohibited by federal law.

The treatments that the new directives clearly specified as not required were the same treatments that the respondents were least likely to think were required on the survey. Perhaps many respondents had understood, even before the distinction was made explicit in the directives, that those who framed the directives had not intended such treatments to be required.

Would the distinction between "futile" and "medically beneficial" treatments in the later versions of the directives lead to consensus concerning which treatments are required? Although there are no survey results based on the new wording of the directives, based on the pattern of responses to the survey, it seems likely that the ambiguity probably would remain. For each case, there was no abrupt division between those treatments for which there was consensus that they should be given, and those for which there was consensus that they should be withheld. Instead, there was a gradual decrease in the percentage of respondents who would recommend each treatment or who thought each treatment was required. This probably was a result of the fact that there is no clear consensus among respondents about whether some treatments would be beneficial or would lead to a futile prolongation of the dying process.

Moreover, it does not seem that the distinction between "medical decisions" and decisions based on "quality of life" considerations would be likely to remove the ambiguity in the interpretation of the directives. Based on the observations in the neonatal intensive care units, and reflected in the pattern of answers on the questionnaire, it is apparent that medical decisions always involve questions of values. Scientific data and methodology alone cannot be used alone to differentiate a "good" choice from a "bad" one. Such differentiations depend on value considerations. Medical decisions about the benefit of treatment involve consideration of the value of extending life in a particular state for a given amount of time. For example, even the recommendations of the framers of the revised directives that certain treatments should be given whereas other may be withheld from anencephalic babies, very premature babies, and those with intracranial bleeds, reflect decisions concerning length and quality of life.[31] Though there might be fewer differences in interpretation, differences in respondents' values would probably lead to a lack of consensus concerning the interpretation of even the revised directives for many cases.

What are the implications of the survey for guidelines that depend on a distinction between ordinary and extraordinary treatment? Because these terms are used in a number of different ways, guidelines that rest on a distinction between ordinary and extraordinary care and do not clearly define and operationalize the terms cannot provide clear standards for treatment decisions. Usage of the terms include distinctions based on (1) how usual or unusual the treatment is, (2) whether the treatment would or would not be beneficial to the particular patient, and (3) on the aggressiveness of the treatment. In addition, there is variation in the categorizations of treatments as ordinary and extraordinary because categorizations according to each of these definitions entail value judgements. *All* treatment choices were rated as falling in the most ordinary category (1) by some respondents and as in the most extraordinary category (5) by others.[32]

The findings of this survey support the conclusion of the President's Commission for the Study of Ethical Problems in Medicine and Biomedical and Behavioral Research, that

> . . . despite its long history of frequent use, the distinction between ordinary and extraordinary treatment has now become so confused that its continued use in the formulation of public policy is no longer desirable.[33]

The distinction between ordinary and extraordinary care could only be a useful part of a guideline for treatment decisions if the meaning was clearly defined.

What are the implications of the survey concerning the manner in which decisions are to be made? Almost all respondents felt that the parents of catastrophically ill newborns should be involved in the decision-making process. Perhaps this is because respondents recognize that medical decisions do involve the beliefs and values of the decision-makers and feel that the parents, as proxy decision-makers for their children, should play a role in decision-making. Respondents were almost evenly split between believing that there should be some general guidelines, at least on the hospital level, to guide decision-makers, and believing that all decisions should be made exclusively on a case-by-case basis.

Conclusions

The results of this survey indicate that, within this knowledgeable group of respondents, there was a high degree of consensus concerning the advisability of recommending some treatments and of withholding others in some cases. However, for many other treatment choices, there was no clear consensus. Likewise, in interpreting the Baby Doe Directives, there was consensus that some treatments would be required, but there was no consensus concerning others. The lack of consensus concerning which treatment would be "futile" for a particular baby and which would be "of benefit" suggest that there would not be consensus about the requirements of the Baby Doe Directives, even as revised in 1984. Since there was much variation in the categorization of some treatments as ordinary or extraordinary, the use of those terms is unlikely to prove useful in the wording of a guideline meant to regulate treatment. Most respondents felt that many treatments would be required by the directives that would not be in the best interests of the babies.

Respondents overwhelmingly rejected the concept of specific federal guidelines to regulate decision-making for catastrophically ill newborns. Nearly half thought that there should be no guidelines and decisions should be made only on a case by case basis, but others felt that there should be general guidelines for decision-making. General guidelines that might recommend some types of treatment in some cases, but that would allow for discretion in making other decisions, would allow the values of the parents and health care professionals to inform decisions based on the particulars of the case, at least for those cases for which there is no clear consensus in our society.

As advances in medical technology and practice continue to take place, new diagnostic procedures will change the information available about patient conditions and new developments will lead to new options for treatment. The prognosis for individual newborns with particular problems will change, but there will continue to be new questions concerning which treatments are best for particular infants. More information will be needed from followup studies on the consequences of particular treatment choices, more information will be needed on the manner in which treatment choices are made, and more discussion and debate will be needed on the humanistic dimensions in the care of imperiled newborns.

Appendix—Text of the Survey[34]

Please read the following four vignettes. For each one assume that:
—The parents' views are the same as yours.
—Each treatment, if given, would increase the baby's chance of survival.
—You are asked to recommend what you think would be the *best* treatment decisions for each baby.

Baby "A"

Baby "A" is born with Down's syndrome (Mongolism). Soon after birth, the baby is also found to have duodenal atresia, an intestinal defect which can be corrected by routine surgery. Without surgery, the baby cannot drink milk or other fluids by mouth.

A. Would you recommend:

	Definitely Yes	Probably Yes	Probably No	Definitely No
1. Intravenous feedings?	1	2	3	4
2. Surgery to correct the intestinal defect?	1	2	3	4
3. Antibiotics, if it is suspected that the baby also had an infection?	1	2	3	4

Suppose the baby was also found to have a heart defect, would you recommend:

	Definitely Yes	Probably Yes	Probably No	Definitely No
4. Cardiac catheterization—an invasive diagnostic procedure?	1	2	3	4
5. Open heart surgery (for VSD)?	1	2	3	4
6. After heart surgery, suppose the baby developed chronic kidney failure, would you recommend maintenance dialysis?	1	2	3	4

The Department of Health and Human Services recently issued a directive stating that hospitals must post signs saying "Discriminatory failure to feed and care for handicapped infants in this facility is prohibited by Federal law." No specific guidelines were issued to aid in interpreting the directive. A decision by a Federal district judge struck down the new rule. The Department of Health and Human Services has appealed the decision.

B. In interpreting the federal directive, do you think the following treatments would be required or not for baby "A"?

	Definitely Yes	Probably Yes	Probably No	Definitely No
1. Intravenous feedings	1	2	3	4
2. Surgery to correct the intestinal defect	1	2	3	4
3. Antibiotics	1	2	3	4
4. Cardiac catheterization	1	2	3	4
5. Open heart surgery	1	2	3	4
6. Kidney dialysis	1	2	3	4

C. On a scale from 1 (ordinary) to 5 (extraordinary), how would you rate the treatments listed above for baby "A"?

	Ordinary		Extraordinary		
1. Intravenous feedings	1	2	3	4	5
2. Surgery to correct the intestinal defect	1	2	3	4	5
3. Antibiotics	1	2	3	4	5
4. Cardiac catheterization	1	2	3	4	5
5. Open heart surgery	1	2	3	4	5
6. Kidney dialysis	1	2	3	4	5

Each of the following vignettes was presented in the same format as the case of Baby "A" with the case presentation followed by questions about treatment recommendations, the requirements of the "Baby Doe Directives," and ratings on the "ordinary/extraordinary scale."

Baby "B"

At birth, Baby "B" is found to be anencephalic (lacking the cerebrum, cerebellum, and the flat bones of the skull), which indicates that the baby could have no upper brain function. Most anencephalic babies die within the first few days of life; all die within the first few weeks.

Treatments:

1. Resuscitation—trying to start respiration if the baby isn't breathing
2. Feeding by mouth if the baby can suck
3. Gavage (tube) feeding if the baby can't suck
4. Antibiotics if it is suspected that the baby also has an infection

Suppose the baby was also found to have a heart defect:

5. Cardiac catheterization—an invasive diagnostic procedure
6. Open heart sugery (for VSD)
7. an arrest page—restarting the heart if it stops beating

Baby "C"

Baby "C" was born with multiple congenital anomalies—low set ears, skin folds around the neck, a cleft palate, and cardiac anomalies—suggestive of trisomy-13, a chromosomal anomaly that is always associated with severe mental retardation and severe physical impairments. Most of these babies die within the first few months, almost all die within the first year. If Baby "C" doesn't have trisomy-13, he may have only correctable physical defects or he may have uncorrectable physical and/or neurological defects.

Treatments:

1. Resuscitation—trying to start respiration—in the delivery room
2. Nutrition and fluids
3. Putting the baby on a respirator if he can't breathe for himself
4. Antibiotics, if it is suspected that the baby has an infection

Now suppose that after resuscitation in the delivery room Baby "C" was breathing on his own and was admitted to the neonatal intensive care unit for evaluation. Two days later, chromosomal analysis indicated that he does indeed have trisomy-13.

5. Nutrition and fluids
6. Surgery to correct the cleft palate

7. Antibiotics, if it is suspected that the baby had an infection
8. Putting the baby on a respirator if he can't breathe for himself
9. Cardiac catheterization
10. Open heart surgery (for VSD)
11. An arrest page—restarting the heart if it stops beating

Baby "D"

Baby "D" was born at a gestational age of 25 weeks (15 weeks before the end of a full-term pregnancy) weighing 560 grams (1 lb 3 oz). He was born vaginally. His Apgar score at birth was 1—a score that indicated the baby had probably suffered from lack of oxygen. His eyes were fused—indicating that he was very premature. He was not breathing on his own, but did have a slow heart beat.

Treatments:

1. Resuscitation—trying to start respiration—in the delivery room

Suppose Baby "D" was resuscitated and put on a respirator: The following day, ultrasound—a diagnostic test—revealed that he had a grade III–IV IVH—a large amount of bleeding in the brain. He therefore has approximately a 50% chance of survival, and if he survives, he probably has less than a 50% chance of being normal. Deficits could range from moderate to severe mental retardation and/or neurological impairments (such as cerebral palsy).

2. Increase respirator settings—giving the baby more oxygen
3. Nutrition and fluids
4. Suctioning to remove excess fluid from airways
5. Pressors—powerful drugs to maintain blood pressure
6. An arrest page to restart the heart
7. Kidney dialysis if the kidneys failed

Acknowledgments

I would like to thank the Conference participants and the members of the staff at Columbia–Presbyterian Medical Center who completed the questionnaire. I also wish to thank John Arras, Arthur Caplan, John Colombotos, Ann Dill, John Driscoll, Zola Golub, Elane Gutterman, L. Stanley Jones, Bruce Levin, Kiyoko Liozzo, Molly Park, Carl Pieper, Katherine Powderly, Katherine Rosasco and the late Raymond Vande Wiele for their help. They each made important contributions to this study.

Notes and References

[1]Parsons, Talcott, Renee C. Fox, and Victor Lidz, "The Gift of Life and Its Reciprocation," *Social Research* **39**:369–415, 1972.

[2]Crane found that there is a growing consensus to treat patients with only physical defects whose lives can be saved, but not to treat those patients who are terminally ill and have mental damage. However, she found there is less consensus concerning treatment for patients whose lives can be saved but have mental damage and those patients who are terminal and have only physical damage. Crane, Diana, *The Sanctity of Social Life: Physicians' Treatment of Critically Ill Patients*, New Brunswick, NJ: Transaction Books, 1977.

[3]Shaw, Anthony, et. al. "Ethical Issues in Pediatric Surgery: A National Survey of Pediatricians and Pediatric Surgeons," *Pediatrics* **60**, no. 4, pt. 2, 588–599 (Oct. 1977).

[4]Reported in the *Hastings Center Report* **6**, no. 4, 3 (April, 1976).

[5]Todres, David, et. al. "Pediatricians' Attitudes Affecting Decision-Making in Defective Newborns," *Pediatrics* **60**, no. 2, 197–201 (Aug. 1977).

[6]Jonsen, A. R., and Garland, M., eds., *Ethics of Newborn Intensive Care*, San Francisco: Health Policy Program, University of California, 1976.

[7]For example, Duff, R. S. and A. G. M. Campbell, "Moral and Ethical Dilemmas in a Special Care Nursery," *New England Journal of Medicine* **289**, 890–894 (1973).

[8]Anspach, Renee, *Life and Death Decisions in Neonatal Intensive Care: A Study in the Sociology of Knowledge*, Berkeley, CA: University of California Press, in press.

[9]*See* John Arras, in this volume.

[10]*See*, for example, Crane, Shaw, et. al., and Todres et. al.

[11]*See* John Arras, in this volume.

[12]See the President's Commission for the Study of Ethical Problems in Medicine and Biomedical and Behavioral Research, *Deciding to Forego Life-Sustaining Treatment*, Washington, DC: The Government Printing Office, 1983 (especially pp. 82–89).

[13]Department of Health and Human Services, Non-discrimination on the Basis of Handicap; Procedures and Guidelines Relating to Health Care for Handicapped Infants, *Federal Register* **49**, no. 8, Thursday, Jan 12, 1984.

[14]It is the conference at which the other papers in this monograph were originally presented.

[15]In addition, four respondents returned questionnaires in both the first and third response groups; their responses were tallied only in the second subsample.

[16]"Consensus and Controversy in Medical Practice", in *Annals AAPSS* **437**, 105 (May, 1978).

[17]Unless otherwise noted, the answers of those who responded "definitely yes" and "probably yes" are combined as are the answers of those who answered "probably no" and "definitely no."

[18]Questions on resuscitation and arrest page for Baby "B" were asked only of the Columbia subsample; there were 117 to 119 responses to each of these questions.

[19]Some would say that a baby like Baby "B" (as well as a baby like Baby "C") lacks those qualities thought to define distinctly human potential, a state they feel has important consequences for decision-making. *See* Arras, this volume.

[20]In later versions of the directives, anencephalic babies were presented as examples of babies who should not receive heroic treatment for such treatment would be futile and would only serve to prolong the dying process. Department of Health and Human Services, Nondiscrimination on the Basis of Handicap Relating to Health Care for Handicapped Infants, *Federal Register* **48**, no. 129, 30846 (Tuesday, July 5, 1983).

[21]All responses for Baby "C" are from the Columbia respondents only; $n = 115$ to 119.

[22]Although respondents were asked directly to rate treatments on the "ordinary/extraordinary scale," the "aggressiveness" of treatments was inferred from the responses to the survey and corresponded to the way that treatments were observed to be ranked in clinical practice.

[23]The transposition score counts the number of transpositions or interchanges of two treatments necessary in order to permute a given subset of treatment choices into a subset of treatment choices which corresponds exactly to the hypothesized ordering of treatments according to aggressiveness. This measure was chosen because it reflects the extent of disagreement from the hypothesized ordering, weighting those subsets more heavily that departed more from the hypothesized norm.

[24]Scores for Baby "C" were available only for the Columbia respondents.

[25]Each time the word "significantly" is used in this section, it refers to a difference in which $p < 0.05$ on a one-tailed t-test. The results in this section are based on the responses of the 48 Catholics, 32 Protestants, 20 Jews, 32 neonatologists, and 75 NICU nurses for whom information was available.

[26]The R^2 for the age only model was 0.097. The R^2 for the regression model with age, sex, professional experience, religion and religiosity, and occupation was 0.15.

[27]Paired t-tests between treatments recommended and thought required were significant for *every* treatment option; $p < 0.05$, for most treatment options $p < 0.01$.

[28]Includes respondents who circled "probably" required, "probably not" required, and "definitely not" required.

[29]Based on 230 responses.

[30]The age-adjusted differences were in the positive direction.

[31]Singer, Peter and Helga Kuhse make the same point in their reply to a letter in *The New York Review of Books,* June 14, 1984, p. 50.

[32]Discrepancies in the categorization of treatments as ordinary and extraordinary were also reported in: Peter Singer, Helga Kuhse, and Cora Singer, "The treatment of newborn infants with major handicaps: a survey of obstetricians and paediatricians in Victoria," *The Medical Journal of Australia,* Sept 17, 1983.

[33]*Deciding to Forego Life-sustaining Treatment,* p. 88; This report contains an excellent discussion of the problems with the use of the ordinary/extraordinary distinction.

[34]Text of the version of the questionnaire distributed to the Columbia respondents.

Conclusion

Arthur L. Caplan

The primary rationale for commissioning and publishing the papers included in this volume was to discover whether those who work in and utilize the techniques of the humanities could be of assistance to health professionals and the general public in addressing the complex moral and value problems arising in contemporary neonatal medicine. Although some of the papers address themselves at least in part to matters of fact or professional attitudes about treatment decisions, the majority of the contributions were written with the aim of being explicitly value-laden and prescriptive.

Can the Humanities Help?

It might appear that the humanities is the last place anyone ought to turn when matters pertaining to medicine are under discussion. After all, at least some would argue, medical matters ought be left to those in the medical profession. But as these papers clearly illustrate, the problems raised by the abilities and skills of contemporary neonatology raise issues of morality, law, and value that cannot possibly be resolved solely by appeals to "the facts."

The real test of the papers in this volume then is whether those in the humanities have delivered on the promise that the methods and approaches of their fields can when heeded provide some assistance to those who must grapple with value problems on a day-to-day basis in the neonatal ICU, or in the home, or in the other institutional settings where children with disabilities or ailments receive care and attention.

207

One criticism frequently leveled against the humanities is that those in the humanities never achieve consensus, are unable or unwilling to recommend specific courses of action, and are far too content to spend their time pondering the abstract, the unusual, or the irrelevant to be of much use to those who are faced with decisions that are both pressing and terrifying.

I believe that the various papers presented in this book give the lie to the charge that those in the humanities cannot or will not apply their disciplines to the resolution of concrete problems in the practical world of medicine. There are, for those who read these papers closely and carefully, a large number of specific suggestions, recommendations, and somewhat surprisingly, endorsements of both belief and behavior in caring for imperiled newborns.

More Than "The Facts"

One subject about which most of the authors are in agreement is that scientizing or medicalizing the treatment of imperiled newborns will not make the moral and value issues disappear. It has always failed to do so in the past; there really are no technical fixes to the kinds of value problems that arise when medicine makes progress in treating its youngest patients. It is simply wrong to believe that medical advances will alleviate or diminish either the degree or frequency with which such problems appear in the future.

Medical success has its economic, legal, and moral price, as recent advances in such areas as organ transplantation, artificial organs, and genetic screening have shown. Neonatology has not been, at least with regard to the valuational aspects of care, any different. For some the hope remains that a technical solution to the vexatious problems of neonatal care are only just around the next corner of biomedical research, but all of the contributors to this collection believe that those who have such hopes will inevitably be disappointed.

The Need for Procedures

There is consensus too about the fact that some set of procedural mechanisms needs to be instituted and maintained to protect infants who cannot look out for their own interests. Whatever assessment is made of the efforts of the Reagan administration or Congress in this regard, members of the health professions, the families of imperiled children, and society must come to some agreement about a method that can be used to protect infants against both the malevolent and benevolent intentions of others who may not always know or be able to articulate a child's interests in an objective manner. Whether it be ethics committees, social service agencies, court-appointed guardians, or some other institutional modality, all of those whose views are represented in this book agree that something must be done to insure that someone is responsible for determining and assuring that the interests of the infant are not cavalierly sacrificed in favor of the interests of others.

The contributors all agree that uniform, rigid bureaucratic policies—however well intended—instituted at the Federal level have not and will not contribute much to the resolution of the value dilemmas that permeate neonatal practice. It is not possible to legislate morality in the neonatal ICUs of America by means of bureaucratic fiat from Washington. Time and time again the contributors note that the moral problems of neonatal care are complex, particularistic and contextual—not the sorts of cases that are well suited to the invocation of a single rule, principle, or policy by parties that lack knowledge about and a sensitivity to the nuances of each particular decision.

Criteria for Decisions in the ICU

The contributors also reached consensus on another point that has been the object of much discussion and debate in both Congress and the media—that some medical conditions are so hopeless as to justify the nontreatment of certain infants. Children who are born dying, children who are born without a brain, and children for whom medical intervention would be, with a high degree of certainty, hopeless should not be made the victims of invasive technology and skills that have been developed with the intention of curing or at least diminishing rather than extending suffering.

The contributors also agree that the suffering and burden that medical treatment may place upon a particular tiny patient, while of paramount importance, are not the only factors that need to be considered by those faced with the painful choice of treating or not treating imperiled newborns. Social policy ought also acknowledge the suffering and burden the treatment of certain imperiled newborns can create for families, professional caregivers and even for society as a whole. Strenuous efforts ought be undertaken to minimize these forms of suffering as well.

Yet another point of consensus can be found; social policy in our society must begin to reflect the fact that there are real alternatives to abortion in thinking about what should be done when handicaps or disabilities are detected early on in a pregnancy. The primary option proposed is adoption, an alternative that, at least according to some contributors to this book, has not been fully explored, examined, thought out or even taken particularly seriously by either government or society.

Nor do those writing from the perspective of the humanities applaud medicine's apparent desire to normalize, to destroy differences, to standardize, to homogenize, to average. These are powerful trends within modern medical practice, trends conditioned by and contributing to the attitude our culture takes to those who seem different or deviant. They are attitudes that run deep, particularly among healthy, successful, well-to-do professionals of the sort who staff hospitals and debate public policies in universities and legislatures. Each of the contributors is concerned to remind the reader about the distorting influences of such values on the policies our society develops for dealing with patients who are in Leslie Fiedler's sense of the term 'freaks'—different and, therefore, in many ways unsettling to us.

Responsibility

Nor is it sufficient to show concern only for neonates in the hospital or intensive care setting. All of us are morally bound to commit resources for the followup care and support of those upon whom we decide to bestow modern medicine's largesse. We have not as a nation fulfilled this obligation.

Finally, every one of the contributors seems to agree that we ought not allow our concern with assuring a minimally decent quality of life for every patient who receives care to obscure the fact that the moral presumption that ought guide health professionals in the course of their everyday work is to save and abet life. Although there are certainly situations in which treatment can only prolong suffering or even prolong dying, and other situations in which so little is known about the outcome of an intervention that the risks may outweigh the hoped for benefits, the moral presumption in favor of life must remain firmly entrenched. The burden of proof must always fall to those who believe that the skills and therapies of the health professions ought not be utilized in caring for a particular infant.

The Need for Continuing Dialog

Ironically, teasing out the many areas of agreement and consensus from all of the papers included here may raise false hopes about the extent to which the humanities can help resolve moral problems in medicine. Those who have carefully read through each of these papers may come away with an overinflated view of what the humanities can or ought be expected to do.

To some extent the consensus represented in this book is misleading because, no doubt, to some extent it is incomplete, representing the views of scholars, who, however diverse their perspectives, cannot represent every shade of opinion. Nevertheless, those who have read and considered each of the papers should be better prepared to debunk some points about which our authors agree and to advance other areas of consensus that the contributors have neglected. The real lasting contribution the humanities can make to understanding and solving value problems in neonatal health care depends ultimately upon continuing and extending the dialog initiated in this volume.

Contributors

John D. Arras, PhD

Montefiore Medical Center

Dr. John D. Arras is the philosopher-in-residence at Montefiore Medical Center, Department of Epidemiology and Social Medicine and a lecturer in community health at the Albert Einstein College of Medicine. He received his doctorate in philosophy from Northwestern University and studied at the School of Law at the University of California at Berkeley. He is a visiting associate professor in the Department of Philosophy, Barnard College. Author of Ethical Issues in Modern Medicine *(Mayfield Publishing Co., 1983). He has written numerous articles on ethical dilemmas in modern medicine.*

Arthur L. Caplan, PhD

The Hastings Center

Dr. Arthur L. Caplan has been an associate for the humanities at The Hastings Center since 1977. He received his doctorate in philosophy at Columbia University. He is the editor and author of numerous books and articles, in the areas of the history of and philosophy of the life sciences, health policy and medical ethics. He has taught medical ethics at Columbia University, College of Physicians and Surgeons.

213

Nancy Neveloff Dubler, LLB

Montefiore Medical Center

Nancy N. Dubler is director of the Division of Legal and Ethical Issues in Health Care in the Department of Social Medicine at Montefiore Medical Center. A graduate of Harvard Law School, Ms. Dubler edits the Journal of Prison Health, Medicine, Law, Corrections, and Ethics. *She is a national representative of the American Public Health Association to the National Coalition for Jail Reform and chairs their Subcommittee on Health Care in Prisons and Jails.*

Leslie A. Fiedler, PhD

State University of New York at Buffalo

Dr. Leslie A. Fiedler, Samuel L. Clemens Professor of Literature at the State University of New York at Buffalo, where he has been a professor of English since 1964 and department chair from 1974 to 1977. Author of more than twenty-five books, including Freaks: Myths and Images of the Secret Self *(Simon & Schuster) 1978, and* Love and Death in the American Novel *(Stein and Day), 1966, Dr. Fiedler has held a Rockefeller Fellowship, two Fulbright Fellowships, the Kenyon Review Fellowship in Criticism, the Christian Gauss Fellowship, and a Guggenheim. He won the Furioso Poetry Prize and was granted an award by the National Institute of Arts and Letters for "excellence in creative writing." Dr. Fiedler earned his doctorate at the University of Wisconsin and did postdoctoral work at Harvard. He has been an associate editor of* Ramparts, The Running Man, Quarterly Review of Film Studies, *and* Studies in Black Literature.

Alan R. Fleischman, MD

Montefiore Medical Center

Dr. Alan R. Fleischman is director of neonatology and associate professor of pediatrics at Montefiore Medical Center and the Albert Einstein College of

Medicine. He is a graduate of Albert Einstein College of Medicine and was an intern and a pediatric resident at Johns Hopkins Hospital. Dr. Fleischman, a member of Phi Beta Kappa and of the Alpha Omega Alpha honor medical society, has written numerous articles and book chapters on the special problems of the fetus and the newborn as well as maternal health during pregnancy.

Joseph F. Kett, PhD

University of Virginia

Dr. Joseph F. Kett is a professor in the department of history at the University of Virginia. His fields are American social and cultural history, especially 19th century; history of education and the history of the family. He is the author of two books: The Formation of the American Medical Profession, 1790–1860; The Role of Institutions *(Yale University Press, 1968), and* Rites of Passage: Adolescence in America, 1790–Present *(Basic Books, 1977). In addition, Dr. Kett has written numerous articles on adolescents, the family, and medical education and practice.*

Betty Wolder Levin, PhD

Columbia University

Dr. Betty Wolder Levin is a medical anthropologist currently doing research on decision making in neonatology. As a member of the Project on Ethics and Values in Health Care at the College of Physicians and Surgeons, Columbia University, she helped to develop curriculum materials and taught bioethics to medical and other health science students. She is presently completing her doctorate in the Division of Sociomedical Sciences at Columbia University and serves as a member of the Neonatal Bioethics Review Committee at Columbia Presbyterian Medical Center.

Ruth Macklin, PhD

Albert Einstein College of Medicine

Dr. Ruth Macklin is an associate professor of bioethics in the Department of Community Health at the Albert Einstein College of Medicine. She was previously an associate for behavioral studies at The Hastings Center, Institute of Society, Ethics and the Life Sciences and Director of the Moral Problems in Medicine Project at Case Western Reserve University. Dr. Macklin's areas of interest include the moral issues involved in human genetics; national health policies; care of the mentally ill and geriatric patients; and informed consent. Dr. Macklin is a member of the Committee on Philosophy and Medicine, American Philosophical Association; a member of the Task Force on Law and Ethics, National Board of Medical Examiners; and serves on the editorial boards of Biomedical Ethics Reviews *and* Einstein Quarterly.

Thomas H. Murray, PhD

Institute for the Medical Humanities, University of Texas Medical Branch—Galveston

Dr. Thomas H. Murray is Associate Professor, Ethics and Public Policy, at the Institute for the Medical Humanities, University of Texas Medical Branch-Galveston. Dr. Murray was formerly an associate for social and behavioral studies at The Hastings Center where he also held a National Endowment for the Humanities fellowship. Prior to that he was National Endowment for the Humanities Fellow at Yale University. Dr. Murray was graduated magna cum laude, President's Scholar, from Temple University and received his doctorate in social psychology from Princeton University. The author of numerous articles in scientific journals and lay publications, Dr. Murray has lectured widely on ethical issues in neonatal care, health policy, and genetic engineering. He is former president of the Association for Integrative Studies and advisory editor of Social Science and Medicine: An International Journal.

David H. Smith, PhD

Indiana University

Dr. Smith is director of the Poynter Center and a professor of religious studies and chairman of the Department of Religious Studies at Indiana University. He is a Fellow of the Hastings Center and serves on the board of directors of the Hospice of Bloomington, Inc. He has written and lectured extensively on the moral and ethical issues surrounding the decision to die and the sanctity of life. He wrote The Achievement of John C. Bennett *(Herder & Herder, 1970), coedited* Love and Society: Essays in the Ethics of Paul Ramsey *(Council on the Study of Religion, Fall, 1974), and was editor of* No Rush to Judgment *(Poynter Center, Indiana University, 1977).*

Margaret O'Brien Steinfels, MA

Christianity and Crisis

Ms. Margaret O'Brien Steinfels is executive editor of Christianity and Crisis *and was formerly the editor of* Hastings Center Report. *Author of* "Who's Minding the Children? The History and Politics of Day Care in America *(Simon and Schuster, 1974), she has written numerous articles on subjects surrounding childbirth, day care, child rearing, and the family dilemmas of this century. Formerly a reporter, columnist, and reviewer for the* National Catholic Reporter, *she received her BA from Loyola University in Chicago, and a master of arts degree in American history from New York University.*

Alan J. Weisbard, JD

Benjamin N. Cardozo School of Law, Yeshiva University

Alan J. Weisbard is an assistant professor at Benjamin N. Cardozo School of Law at Yeshiva University and visiting assistant professor of community health at Albert Einstein College of Medicine. A graduate of Harvard College and Yale Law School, Professor Weisbard was assistant director for legal studies to the President's Commission on Ethics in Medicine and Research. His areas of interest include issues surrounding bioethics and the law, and legal issues concerning children and families.

Index

219

DATE DUE

GAYLORD   PRINTED IN U.S.A.